# 新楼盘
## NEWHOUSE  图解地产与设计

49

中西部楼盘

中国林业出版社

旭阳国际商城
项目地点：新疆伊犁
建筑面积：16.5万 平方米

**上海中建建築設計院有限公司**
SHANGHAI ZHONGJIAN ARCHITECTURAL DESIGN INSTITUTE CO., LTD.

**上海中建 · 上海总部**
浦东新区东方路989号
中 达 广 场 12 楼
电话：86 - 21 - 6875 8810

**上海中建 · 西安分院**
南二环西段202号
九 座 花 园 611 室
电话：86 - 29 - 8837 8506

上海中建建筑设计院有限公司，成立于1984年，隶属于中国建筑工程总公司，是国家批准的拥有建筑工程设计甲级、装饰设计甲级、风景园林设计乙级、城市规划编制乙级资质的综合性设计公司。公司是中国勘察设计协会和上海市勘察设计协会会员单位，2011年被授予首批"全国诚信单位"，并入选为中国2010年上海世博会建筑设计类九家推荐服务供应商之一。

公司长期面向海内外开展设计业务，先后与美国、加拿大、法国、比利时、香港等国家和地区的著名设计公司合作设计了多项工程，设计作品遍及国内二十多个省市自治区及海外的俄罗斯、阿尔及利亚等国家，有多项设计作品荣获建设部、上海市及中国建筑工程总公司的各类奖项。

公司将本着敬业、诚信、协作、创新的企业精神，恪守诚信原则，聚焦客户需求，为客户价值的提升呈现专业服务，为人居环境的改善描绘美好蓝图。

## 上海中建设计

中泰雅居
项目地点：新疆乌鲁木齐
建筑面积：66.85万 平方米

**上海中建·新疆分院**
乌鲁木齐市 南湖南路 145 号
上海市政府驻疆办事处 205 室
电 话：86 - 991 - 4161 028

**上海中建·济南分院**
市 中 区 经 四 路 288号
恒 昌 大 厦 1503 室
电 话：86 - 531 - 80992217

**上海中建·安徽分院**
合肥市 包河区 马鞍山南路
绿地赢海大厦 C座 1110室
电 话：86 - 551 - 3711 566

**上海中建·厦门分院**
厦门市 软件园二期 观日路 601
电 话：86 - 592 - 5999 928

### 誠聘精英
WWW.SHZJY.COM

徐州市民活动中心

东莞人民医院新院

夜晚
这里成为宁静、放飞梦想的天堂
金中海·蓝钻

龙跃·新世纪广场

# 深 圳 市 東 大 建 築 設 計 有 限 公 司
## Shenzhen Dongda Architectual Design.,Ltd

### 東南大學建築設計研究院有限公司深圳分公司
Architectural Design & Research, Southeast University, Shenzhen Branch.,Ltd

　　深圳市东大建筑设计有限公司，建筑工程综合甲级设计机构，以东南大学雄厚的建筑学科与人才优势为背景，20多年深圳设计之都的历练，在建筑界享有盛誉的"东大设计"品牌。专业的建筑设计服务涵盖了前期策划咨询、规划设计、建筑工程设计、医疗专项设计统筹、室内及景观设计等建筑全过程设计。

　　"东大设计"现有员工150余人，拥有众多的技术人才，丰富的实践经验，严格的管理制度，完善的质保体系。致力于成为深圳乃至国内一流的设计服务公司，为顾客提供最专业和更优质的服务。

**ARCHITECTURE**
建筑设计

**PLANNING**
总体规划

**INTERIOR DESIGN**
室内设计

**LANDSCAPE DESIGN**
景观设计

地址：深圳市深南中路6031号杭钢富春商务大厦8楼
add: No.6031, 8th Floor, Fuchun Commercial Building, Shennan Road, Shenzhen, China
e-mail: ddfy_szb@21cn.net
Tel.：0755-82999545
fax ：0755-82995667
www.seusz.com

易思博软件基地

左一为主设计师赖聚奎教授为习主席作介绍　　　　　　　　　　　　北京人民大会堂 - 福建庭

# SJS 四季园林

湖南郴州. 龙泉铭邸

■ 风景园林专项设计乙级
□ 景观设计　　　　Landscape Design
□ 旅游建筑设计　　Tour Architectural Design
□ 旅游度假区规划　Resorts and Leisure Planning
□ 市政公园规划　　Park and Green Space Planning

广州市四季园林设计工程有限公司成立于2002年，公司由创始初期从事景观设计，已发展为旅游区规划、度假区规划、度假酒店、旅游建筑、市政公园规划等多类型设计的综合性景观公司。设计与实践相结合，形成了专业的团队和服务机构，诚邀各专业人士加盟。

Add： 广州市天河区龙怡路117号银汇大厦2505
Tell： 020-38273170　　　Fax： 020-86682658
E-mail:yuangreen@163.com　Http://WWW.gz-siji.com

英国ALT建筑景观设计有限公司
深圳市雅蓝图景观工程设计有限公司

| 总体规划 | 住宅空间 | 公共空间 | 旅游度假 |
| MASTER PLAN | RESIDEMTIAL SPACE | PUBLIC SPACE | RESORT |

本期为 南昌金域蓝湾景观设计 项目

**英国 ALT 建筑景观设计有限公司**
**深圳市雅蓝图景观工程设计有限公司** 作为国内最具专业影响力的建筑景观及设计公司之一，在英国、泰国、中国深圳、上海均设有分支机构。专业从事大型综合性公共绿地开发规划，城市规划和城市设计，旧城改造，度假酒店，住宅社区和商业开发项目的园林景观设计以及风景区与主题公园规划。

雅蓝图设计经过多年的磨砺，始终坚持倡导其"国际化视野，地域化实践，设计提升价值"的设计理念，从国内市场与国际水准相结合的层面进行运作，并与国际著名设计事务所建立长期合作伙伴关系。凭借着对于本土市场的深刻认识，以西方现代设计理念与中国本土文化相结合，为客户提供系统化、全程化、国际化的全方位专业设计服务。

我们将充分关爱地球环境和景观，并持续不断探索和研究，以国际化的视野，现代的设计理念，充分尊重人性的需求，注重本土文化的结合，力求通过景观的营造为社会创造美好的人居环境，为客户创造更大的价值，为生态环境作出更大的贡献。

**雅蓝图景观为适应公司不断地发展，**
**长期招募专业的景观设计人才能加入我们这个团队！**

深圳市 南山区 南海大道2009号 新能源大厦 A座 6D
电话：(+86 755) 26650631 26650632　传真：(+86 755) 26650623
邮箱：yalantu@163.com　project@yalantu.com（项目）　admin@yalantu.com（行政）
公司QQ：355010039　邮编：518054
网址： http://www.yalantu.comv

JOIN CHINA INTERNATIONAL
CONCEPT DESIGN AND CONSULTANT INC.

**核心竞争力**
产品定位／经济分析／产品创新／细部节点及材料控制／后期配合服务／质量管理体系

**合作伙伴**
复地集团／金科集团／融汇地产／招商地产／龙湖地产／融侨集团／中海地产

**虚位以待**
重庆公司期待专业人士加入：总建筑师／设计总监／项目建筑师／主创建筑师／建筑师／助理建筑师／实习建筑师

# GVL 怡境景观
GREENVIEW LANDSCAPE DESIGN LIMITED

- 景观设计 — Landscape Design
- 旅游度假项目规划 — Resorts and Leisure Planning
- 市政项目规划 — Urban Planning Design
- 居住环境项目规划 — Community Planning
- 公园及娱乐项目规划 — Parks and Entertainment Planning and Design

英国国家园景
工业协会海外会员

British Association of Landscape Industries,
Overseas Full Member of BALI

美国景观设计师协会企业会员

Corporate Member Of American
Society Of Landscape Architects

**广州总部**
地　址：珠江新城华夏路49号津滨腾越大厦南塔8楼
邮　编：510623
电　话：(020) 87690558　　87695498
　　　　　38032762　　38032729
传　真：(020) 87697706
邮　箱：greenv@163.com
网　址：www.greenview.com.cn

**香港**
地　址：北角渣华道18号嘉汇商业大厦2106室
邮箱编号：070432
传　真：(852) 22934388

**北京**
地　址：朝阳区亚运村阳光广场B2-1701室
邮　编：100101
传　真：(010) 64977992
电　话：(010) 64975897

## www.greenview.com.cn

@GVL怡境景观
http://weibo.com/gvlcn

www.greenview.com.cn

前言 EDITOR'S NOTE

# 浅谈地产界的"中西部崛起"
## "RISE OF CENTRAL AND WESTERN CHINA" IN REAL ESTATE INDUSTRY

从目前来看，房地产市场的整体盘升已是不争的事实，而其中又以环渤海、长三角、珠三角为轴心的板块轮动作为发展脉络，接下来将是其他二、三线城市和西部地区。受板块轮动效应波及的影响，中西部地区的崛起也不容忽视，除了中央扶持中西部崛起的政策优势外，相比一线城市低廉得多的地价与更高的利润，也是目前各路发展商进入这一地区的主因。从"城市化"到"城镇化"的转变，悄然改变着中西部城市的发展空间。

中西部楼市的崛起受益于轨道交通与高铁建设的跃进，区域规划、城镇化加速，以及一线城市房价过高、产业转移，二、三线城市房价收入比低等一系列因素，中西部城市已然成为房企的又一"主战场"。这些楼盘在空间设计上更加注重整体的和谐，以挖掘每个城市的不同魅力。在整体的设计定位上，既要保持城市独特的魅力，又需兼备时代标准的品质。相比一线城市，二、三线城市更是一种留守城市，是许多人记忆中的故乡，其丰厚的历史积淀出一种特有的人文气息。具体设计方面，如功能区位、历史文化风貌、自然生态环境、建筑造型设计、色彩选择、风格定位等，都旨在营造出这类城市安静、闲适的生活氛围。

建筑是活的艺术，中西部楼市的发展也将会如一片清新的蓝海，为人们建造生态宜居的乐园。本期专题精选部分中西部楼盘典范案例，邀您一同探讨。

Seen from today's market, the continuing warming of the real estate industry is obvious. Based in Bohai Sea District, Yangtze River Delta and Pearl River Delta, real estate development will expand to the second and third-tier cities as well as the western Chinese cities. The rise of central and western China is remarkable in recent years. Except the supportive policies by the government, low land price and high profit are the main reasons to attract developers. The cities in the central and western China have developed a lot during the process of urbanization.

Real estate development in the central and western China benefits a lot from the construction of railways. Due to the rapid urbanization and the high house price in the first-tier cities, some real estate enterprises begin to expand their business to the cities in central and western China. Space design for the housing projects in these regions pays more attention to integrity and tries to highlight their regional characteristics. In terms of the overall design, it should not only keep the uniqueness of the city but also reach to modern quality standard. Comparing to the first-tier cities, second and third-tier cities are hometowns for many people, embodying different local cultures. Detail designs such as the function design, cultural design, environmental design, architectural design, color design and style design should highlight the peaceful and cozy atmosphere of these cities.

Architecture is a kind of living art, thus the real estate development in central and western China will bring people with ecological and comfortable homes. In this issue, we've selected some typical projects and invited you to study on them with us.

jiatu@foxmail.com

**新楼盘** NEWHOUSE 图解地产与设计

2013年 总第49期

面向全国上万家地产商决策层、设计院、建筑商、材料商、专业服务商的精准发行

**指导单位 INSTRUCTION UNIT**
亚太地产研究中心
中国花卉园艺与园林绿化行业协会

**出品人 PUBLISHER**
杨小燕 YANG XIAOYAN

**主编 CHIEF EDITOR**
王 志 WANG ZHI

**副主编 ASSOCIATE EDITOR**
熊 冕 XIONG MIAN

**编辑记者 EDITOR REPORTERS**
唐秋琳 TANG QIULIN
钟梅英 ZHONG MEIYING
胡明俊 HU MINGJUN
康小平 KANG XIAOPING
严琪琪 YAN QIQI
黄洁桦 HUANG JIEHUA
吴 辉 WU HUI
曾伊莎 ZENG YISHA
曹丹莉 CAO DANLI
王盼青 WANG PANQING

**设计总监 ART DIRECTORS**
杨先周 YANG XIANZHOU
何其梅 HE QIMEI

**美术编辑 ART EDITOR**
詹婷婷 ZHAN TINGTING

**国内推广 DOMESTIC PROMOTION**
广州佳图文化传播有限公司

**市场总监 MARKET MANAGER**
周中一 ZHOU ZHONGYI

**市场部 MARKETING DEPARTMENT**
方立平 FANG LIPING
熊 光 XIONG GUANG
王 迎 WANG YING
杨先凤 YANG XIANFENG
熊 灿 XIONG CAN
刘 佳 LIU JIA
王成林 WANG CHENGLIN
刘 能 LIU NENG

图书在版编目（CIP）数据
新楼盘. 中西部楼盘：汉英对照 / 佳图文化主编. ——北京：中国林业出版社, 2013.5
ISBN 978-7-5038-7037-8

Ⅰ.①新... Ⅱ.①佳... Ⅲ.①建筑设计 - 中国 - 现代 - 图集 Ⅳ.①TU206

中国版本图书馆CIP数据核字(2013)第091901号
出版：中国林业出版社
主编：佳图文化
责任编辑：李顺 许琳
印刷：利丰雅高印刷（深圳）有限公司

**特邀顾问专家 SPECIAL EXPERTS** (排名不分先后)

| | |
|---|---|
| 赵红红 ZHAO HONGHONG | 林世彤 LIN SHITONG |
| 王向荣 WANG XIANGRONG | 熊 冕 XIONG MIAN |
| 陈世民 CHEN SHIMIN | 周 原 ZHOU YUAN |
| 陈跃中 CHEN YUEZHONG | 李焯忠 LI ZHUOZHONG |
| 邓 明 DENG MING | 原帅让 YUAN SHUAIRANG |
| 冼剑雄 XIAN JIANXIONG | 王 颖 WANG YING |
| 陈宏良 CHEN HONGLIANG | 周 敏 ZHOU MIN |
| 胡海波 HU HAIBO | 王志强 WANG ZHIQIANG / DAVID BEDJAI |
| 程大鹏 CHENG DAPENG | 陈英梅 CHEN YINGMEI |
| 范 强 FAN QIANG | 吴应忠 WU YINGZHONG |
| 白祖华 BAI ZUHUA | 曾繁柏 ZENG FANBO |
| 杨承刚 YANG CHENGGANG | 朱黎青 ZHU LIQING |
| 黄宇奘 HUANG YUZANG | 曹一勇 CAO YIYONG |
| 梅 坚 MEI JIAN | 冀 峰 JI FENG |
| 陈 亮 CHEN LIANG | 滕赛岚 TENG SAILAN |
| 张 朴 ZHANG PU | 王 毅 WANG YI |
| 盛宇宏 SHENG YUHONG | 陆 强 LU QIANG |
| 范文峰 FAN WENFENG | 徐 峰 XU FENG |
| 彭 涛 PENG TAO | 张奕和 EDWARD Y. ZHANG |
| 徐农思 XU NONGSI | 郑竞晖 ZHENG JINGHUI |
| 田 兵 TIAN BING | 刘海东 LIU HAIDONG |
| 仇益国 QIU YIGUO | 凌 敏 LING MIN |
| 李宝章 LI BAOZHANG | 谢锐何 XIE RUIHE |
| 李方悦 LI FANGYUE | 姜 圣 JIANG SHENG |
| 林 毅 LIN YI | 章 强 ZHANG QIANG |
| 陈 航 CHEN HANG | 田守能 TIAN SHOUNENG |
| 范 勇 FAN YONG | 袁 凌 YUAN LING |
| 赵士超 ZHAO SHICHAO | 满 志 MAN ZHI |
| 孙 虎 SUN HU | 孙明炜 SUN MINGWEI |
| 梅卫平 MEI WEIPING | |

编辑部地址：广州市海珠区新港西路3号银华大厦4楼
电话：020—89090386/42/49、28905912
传真：020—89091650

北京办：王府井大街277号好友写字楼2416
电话：010—65266908  传真：010—65266908

深圳办：深圳市福田区金田路福岗园2412
电话：0755-83586026 /82538737  传真：0755—82538737

**协办单位 CO—ORGANIZER**

广州市金冕建筑设计有限公司 熊冕 总设计师
地址：广州市天河区珠江西路5号国际金融中心主塔21楼06—08单元
TEL：020—88832190  88832191
http://www.kingmade.com

**AECF** 上海颐朗建筑设计咨询有限公司 巴学天 上海区总经理
地址：上海市杨浦区大连路970号1308室
TEL：021—65909515  FAX：021—65909526
http://www.yl—aecf.com

深圳市东大建筑设计有限公司
地址：深圳市深南中路6031号杭钢富春商务大厦8楼
TEL：0755-82999545  FAX：0755-82995667
http://www.seusz.com

**WEBSITE COOPERATION MEDIA**
网站合作媒体

SouFun 搜房网    ABBS 建筑论坛 .com.cn Architecture BBS

## 副理事长单位 DEPUTY CHAIRMAN

华森建筑与工程设计顾问有限公司　邓明　广州公司总经理
地址：深圳市南山区滨海之窗办公楼6层
　　　广州市越秀区德政北路538号达信大厦26楼
TEL：0755—86126888　020—83276688
http://www.huasen.com.cn　E—mail:hsgzaa@21cn.net

广州瀚华建筑设计有限公司　冼剑雄　董事长
地址：广州市天河区黄埔大道中311号羊城创意产业园2—21栋
TEL：020—38031268　FAX：020—38031269
http://www.hanhua.cn
E—mail：hanhua—design@21cn.net

上海中建建筑设计院有限公司　徐峰　董事长
地址：上海市浦东新区东方路989号中达广场12楼
TEL：021—68758810　FAX：021—68758813
http://www.shzjy.com
E—mail：csaa@shzjy.com

## 常务理事单位 EXECUTIVE DIRECTOR OF UNIT

深圳市华域普风设计有限公司　梅坚　执行董事
地址：深圳市南山区海德三道海岸城东座1306—1310
TEL：0755—86290985　FAX：0755—86290409
http://www.pofart.com

天萌（中国）建筑设计机构　陈宏良　总建筑师
地址：广州市天河区员村四横路128号红专厂F9栋天萌建筑馆
TEL：020—37857429　FAX：020—37857590
http://www.teamer—arch.com

GVL国际怡境景观设计有限公司　彭涛　中国区董事长及设计总监
地址：广州市珠江新城华夏路49号津滨腾越大厦南塔8楼
TEL：020—87690558　FAX：020—87697706
http://www.greenview.com.cn

奥雅设计集团　李宝章　首席设计师
深圳总部地址：深圳蛇口南海意库5栋302
TEL：0755—26826690　FAX：0755—26826694
http://www.aoya—hk.com

R—LAND北京源树景观规划设计事务所　白祖华　所长
地址：北京朝阳区朝外大街怡景园5—9B
TEL：010—85626992/3　FAX：010—85625520
http://www.ys—chn.com

广州山水比德景观设计有限公司　孙虎　董事总经理兼首席设计师
地址：广州市天河区珠江新城临江大道685号红专厂F19
TEL：020—37039822/823/825　FAX：020—37039770
http://www.gz—spi.com

北京寰亚国际建筑设计有限公司　赵士超　董事长
地址：北京市海淀区西四环北路15号依斯特大厦102
TEL：010—65797775　FAX：010—84682075
http://www.hygjjz.com

广州市四季园林设计工程有限公司　原帅让　总经理兼设计总监
地址：广州市天河区龙怡路117号银汇大厦2505
TEL：020—38273170　FAX：020—86682658
http://www.gz—siji.com

奥森国际景观规划设计有限公司　李焯忠　董事长
地址：深圳市南山区南海大道1061号喜士登大厦四楼
TEL：0755—26828246　86275795　FAX：0755—26822543
http://www.oc—la.com

深圳市佰邦建筑设计顾问有限公司　迟春儒　总经理
地址：深圳市南山区兴工路8号美年广场1栋804
TEL：0755—86229594　FAX：0755—86229559
http://www.pba—arch.com

深圳市雅蓝图景观工程设计有限公司　周敏　设计董事
地址：深圳市南山区南海大道2009号新能源大厦A座6D
TEL：0755—26650631/26650632　FAX：0755—26650623
http://www.yalantu.com

北京博地澜屋建筑规划设计有限公司　曹一勇　总设计师
地址：北京市海淀区中关村大街31号神舟大厦8层
TEL：010—68118690　FAX：010—68118691
http://www.buildinglife.com.cn

北京新纪元建筑工程设计有限公司　曾繁柏　董事长
地址：北京市海淀区小马厂6号华天大厦20层
TEL：010—63483388　FAX：010—63265003
http://www.bjxinjiyuan.com

香港华艺设计顾问（深圳）有限公司　林毅　总建筑师
地址：深圳市福田区华富路都大厦14、15楼
TEL：0755—83790262　FAX：0755—83790289
http://www.huayidesign.com

HPA上海海波建筑设计事务所　陈立波、吴海青　公司合伙人
地址：上海中山西路1279弄6号楼国峰科技大厦11层
TEL：021—51168290　FAX：021—51168240
http://www.hpa.cn

哲思（广州）建筑设计咨询有限公司　郑竟晖　总经理
地址：广州市天河区天河北路626号保利中宇广场A栋1001
TEL：020—38823593　FAX：020—38823598
http://www.zenx.com.au

深圳文科园林股份有限公司　李从文　董事长
地址：深圳福田区滨河大道中央西谷大厦21层
TEL：0755—36992000　FAX：0755—83148398
http://www.wksjy.com

深圳禾力美景规划与景观工程有限公司　袁凌　设计总监
地址：深圳市福田区泰然九路红松大厦B座9G
TEL：0755-82988000　FAX：0755-83933215
http://www.wlklandscape.com

## 理事单位 COUNCIL MEMBERS （排名不分先后）

广州市柏澳景观设计有限公司　徐农思　总经理
地址：广州市天河区广园东路2191号时代新世界中心南塔2704室
TEL：020—87569202　FAX：020—87635002
http://www.bacdesign.com.cn

中房集团建筑设计有限公司　范强　总经理／总建筑师
地址：北京市海淀区百万庄建设部院内
TEL：010—68347818

北京奥思得建筑有限公司　杨承冈　董事总经理
地址：北京朝阳区东三环中路39号建外SOHO16号楼2903—2905
TEL：86—10—58692509/19/39　FAX：86—10—58692523

陈世民建筑师事务所有限公司　陈世民　董事长
地址：深圳市福田中心区益田路4068号卓越时代广场4楼
TEL：0755—88262516/429

广州嘉柯园林景观设计有限公司　陈航　执行董事
地址：广州市珠江新城华夏路49号津滨腾越大厦北塔506—507座
TEL：020—38032521/23　FAX：020—38032679
http://www.jacc—hk.com

侨恩国际（美国）建筑设计咨询有限公司
地址：重庆市渝北区龙湖MOCO4栋20—5
TEL：023—88197325　FAX：023—88197323
http://www.jnc—china.com

CDG国际设计机构　林世彤　董事长
地址：北京海淀区长春路11号万柳亿城中心A座10/13层
TEL：010—58815603　58815633　FAX：010—58815637
http://www.cdgcanada.com

广州市圆美环境艺术设计有限公司　陈英梅　设计总监
地址：广州市海珠区宝岗大道杏坛大街56号二层之五
TEL：020—34267226　83628481　FAX：020—34267226
http://www.gzyuanmei.com

上海唯美景观设计工程有限公司　朱黎青　董事、总经理
地址：上海徐虹中路20号2—202室
TEL：021—61122209　FAX：021—61139033
http://www.wemechina.com

上海金创源建筑设计事务所有限公司　王毅　总建筑师
地址：上海市杨浦区黄兴路1858号701—703室
TEL：021—55062106　FAX：021—55062106—807
http://www.odci.com.cn

深圳灵顿建筑景观设计有限公司　刘海东　董事长
地址：深圳福田区红荔路花卉世界313号
TEL：0755—86210770　FAX：0755—86210772
http://www.szld2005.com

深圳市奥德景观规划设计有限公司　凌敏　董事总经理、首席设计师
地址：深圳市南山区蛇口海上世界南海意库2栋410#
TEL：0755—86270761　FAX：0755—86270762
http://www.lucas—designgroup.com

广州邦景园林绿化设计有限公司　谢锐何　董事及设计总监
地址：广州市天河北路175号祥龙花园晖祥阁2504/05
TEL：020—87510037　FAX：020—38468069
http://www.bonjinglandscape.com

上海天隐建筑设计有限公司　陈锐　执行董事
地址：上海市杨浦区国康路100号国际设计中心1402室
TEL：021-65988000　FAX：021—65982798
http://www.skyarchdesign.com

目录 CONTENTS

020

058

011 前言 EDITOR'S NOTE

016 资讯 INFORMATION

### 名家名盘 MASTER AND MASTERPIECE

020 上海绿地嘉尚国际：体现简约古典风格的高端商业空间
UPSCALE COMMERCIAL SPACE WITH SIMPLE AND CLASSICAL STYLE

028 南通山水一号：张弛有度的草原式田园风格社区
COMMUNITY IN PRAIRIE AND PASTORAL STYLE WITH FLEXIBLE DEGREE

### 专访 INTERVIEW

038 商业综合体的设计要做好业态整合与因地制宜
——访HMA董事、设计总监 东英树
THE DESIGN OF COMMERCIAL COMPLEX SHOULD PAY ATTENTION TO THE INDUSTRY INTEGRATION AND ACT ACCORDING TO THE LOCAL CONDITIONS

042 景观设计要注重差异化的空间营造
——访广州邦景园林绿化设计有限公司董事、设计总监 谢锐何
DIFFERENTIATION OF THE SPACE IN LANDSCAPE DESIGN

### 新景观 NEW LANDSCAPE

046 佛山南海天安数码城四期：展现时尚与科技元素的现代简约景观空间
MODERN SIMPLE LANDSCAPE SPACE WITH FASHION AND TECHNOLOGY ELEMENTS

052　深圳太古城花园：古典园林的现代演绎
CLASSICAL GARDEN WITH MODERN ELEMENTS

058　庐山云岭：掬一捧地中海风情 钓一串高尔夫时光
MEDITERRANEAN STYLE AND GOLF ENJOYMENT

## 专题 FEATURE

068　呼和浩特秋实璟峯汇：
绿色、适用、经济型的中高标准居住社区
ECOLOGICAL, AFFORDABLE AND ECONOMICAL UPPER-MIDDLE STANDARD RESIDENTIAL COMMUNITY

076　成都南郡七英里：中西结合畅享都市内的庄苑生活
URBAN GARDEN LIFE COMBINING CHINESE AND WESTERN ELEMENTS

084　重庆隆鑫鸿府：新巴渝风的花园式居住社区
GARDEN RESIDENTIAL COMMUNITY IN NEW BAYU (CHONGQING) STYLE

094　成都三盛·翡俪山：原生资源，造就成都第一纯别墅区
NATURAL RESOURCES MAKING FIRST VILLA COMMUNITY IN CHENGDU

## 新特色 NEW CHARACTERISTICS

104　深圳和亨城市广场：尽享巍峨山景与浪漫海景的高档住宅区
UPSCALE RESIDENTIAL COMMUNITY WITH SPECTACULAR MOUNTAIN VIEWS AND ROMANTIC SEASCAPE

## 新空间 NEW SPACE

112　《爱马士之家》置信联排别墅-中间户型：高贵 时尚 典雅 自然
LUXURY, FASHION, ELEGANCE, NATURE

## 新创意 NEW IDEA

118　博斯普鲁斯新城：强调绿色空间的滨水住宅
WATERSIDE RESIDENCES THAT HIGHLIGHT GREEN SPACE

# 商业地产
# COMMERCIAL BUILDINGS

126　卡伦艺术系列酒店："凸" "凹"有致的"绿色"建筑
GREEN BUILDING WITH CANTILEVERED BOXES AND RANDOM POCKETS

132　武汉融科·天城：融水之灵性、流畅、自然于一体的"城市客厅"
"URBAN SITTING ROOM" WITH FLEXIBILITY, SMOOTHNESS AND NATURE

140　温特和克希尔顿酒店：蕴含艺术与意象之美的城市地标
A LANDMARK WITH ART AND IMAGERY

# INFORMATION  |  资讯/地产

## 2012上市房企年报大盘点

3月中下旬，上市房企进入年报密集发布期。从年报可以看出，2012年仍是房地产市场的牛市：万科以1412.3亿元创下销售金额新高，中海以128亿元的净利润创下房企盈利新高，恒大则创下销售面积新高。但在规模扩大、负债率降低的同时，行业的利润率持续下跌。

### LISTED REAL ESTATE ENTERPRISES RELEASED 2012 ANNUAL REPORTS

In the second half of March, the listed real estate enterprises have released their annual datas for the sales in 2012. Known from these reports, 2012 is still a bull market for real estate sales. Vanke's annual sales hit a record high of RMB 141.23 billion, CR Land made the highest net profits of RMB12.8 billion, while Evergrande ranked No.1 in term of single-month sales area for nine consecutive months in 2012. However, with the expanding of the scale and the reducing of the debt ratio, the rate of profit for real estate industry has reduced consecutively.

## 万达将签约广州四大项目总投资450亿

近期，万达集团在广州表示，将斥资逾450亿元在广州再建4个大型项目，其中包括3个万达广场和1个总投资额超过200亿元的万达旅游文化城项目，并于近日完成上述4个项目的签约。4个全新项目签约后，万达集团在广州将有7个项目。

### WANDA TO INVEST 45 BILLION IN GUANGZHOU TO BUILD FOUR NEW PROJECTS

Wanda Group will invest over 45 billion yuan ($ 7.2 billion) to build four more large projects in Guangzhou. The projects include three Wanda Plazas, and Wanda City of Culture and Tourism with an investment of 20 billion yuan ($3.2 billion). The contracts for the above projects were signed. Together with these four new projects, Wanda Group now has totally 7 projects in Guangzhou.

## 江西农民建房用上专家特色设计图

近期，数万套《江西省和谐秀美乡村特色农房设计图集》已免费供农民建房使用。《图集》共分富裕型、小康型、经济型三类户型。江西省住建厅集中半年多时间，组织全省20多家设计院所的专家，通过两轮设计方案的优化和专家评审，从50个农房设计方案中精选出30个优秀方案，汇编成集。江西省住建厅还将《图集》批准为江西省地方标准，目前正在组织推广应用。

### JIANGXI FARMERS: BUILD HOUSE WITH PROFESSIONAL ARCHITECTURAL DESIGN

Recently, thousands of Collection of Rural Housing Design Drawings for Jiangxi Province are provided to the farmers free. The collection includes drawings for three types of houses: luxury type, middle-class type and economical type. For a half year, Bureau of Housing and Urban-Rural Development of Jiangxi Province (BHURD) has organized more than 20 experts of design institutes to select 30 excellent housing designs from 50 concepts and composed this collection. Now the Collection is approved by the BHURD of Jiangxi Province and will be promoted widely.

## 华润1.6亿竞得烟台莱山区地块 计划建综合体

近期，华润集团旗下华润置地正式入驻烟台，竞得莱山区观海路商住用地项目。据悉，观海路项目位于莱山城区的核心，总建筑面积近100万m²，该项目将打造成涵盖高端购物中心万象城、国际五星级度假酒店、国际5A级写字楼、高档公寓和高尚精品住宅的多业态现代都市综合体。

### CR LAND: 160 MILLION WON LAISHAN PLOT FOR COMMERCIAL COMPLEX

CR Land under China Resources Group has won the plot on Guanhai Road, Laishan District of Yantai City. The site is in the center of downtown Laishan with a total floor area of 1,000,000 m². The project is designated to be an urban complex including the MIXC, five-star resort hotel, grade-A office building, high-end apartments and boutique residences.

## 武汉：第10届园博园主题建筑规划公布

近日，记者从武汉市政府召开的第十届中国(武汉)国际园林博览会新闻发布会上获悉，园博园建设已有序展开，四大主题建筑规划首次露脸，明年基本完成建园布展。此次园博园总体规划突出生态、科技、人文、民生、地域五大特色，结合现状地形、地貌及本届园博会的特点，以"山水连枝，双轴两区，一心三馆"为整体园区规划结构。

### THEME BUILDING PLANNING FOR THE 10TH CHINA (WUHAN) INTERNATIONAL GARDEN EXPO ANNOUNCED

According to Wuhan government's press release for the 10th China (Wuhan) International Garden Expo Announced, the construction of the garden has begun. The planning for the four theme buildings was first announced, and it plans to complete the construction within next year. The master plan highlights ecology, science and technology, culture, livelihood and regions. According to different geography, landform and the theme of this expo, the garden is designed with "one center, three pavilions, two axes and two zones".

## 广州国五条细则出台 补税补社保购房双双叫停

广州市国五条细则通过官方微博正式对外发布，内容要求广州限购将进一步从严，规定"非本市户籍居民家庭持补缴纳税证明或社保证明购房的，不予认可等"；而对于二手房交易如何征税尚未有明确说法，官方表示将联手地税局研究后再对外公布。

### GUANGZHOU ANNOUNCES DETAILED PROPERTY REGULATION MEASURES

Guangzhou announced detailed property regulation measures on its government's official website, and according to the new regulations, Guangzhou needs to continue to strictly implement the existing restrictions that apply to the purchase of residential property. Those without hukou in the city are permitted to buy houses after they have continually paid tax or social security fees there for one year or longer two years before they buy a house. However, authorities did not explain how they will implement the 20-percent tax on capital gains from property sales and will do so when doing research with the local tax bureau.

## 北京国五条细则发布：京籍单身限购一套住房

北京楼市国五条调控细则规定自3月31日起禁止京籍单身人士购买二套房；严格按个人转让住房所得的20%征收个人所得税，出售五年以上唯一住房免征个税；进一步提高二套房贷首付款比例，新旧政策以存量房买卖合同网签时间为准。

### BEIJING ANNOUNCES PROPERTY REGULATION DETAILS: SINGLE ADULTS WITH THE CAPITAL'S HUKOU ALLOWED TO PURCHASE ONLY ONE APARTMENT

Beijing's detailed property regulations state that, from March 31, single adults with the capital's hukou – registered permanent residence – are allowed to purchase only one apartment. Beijing said it will strictly implement the 20-percent tax on capital gains from property sales, but the tax will be exempt if the property is the seller's only one and they have owned it for five years or more. Meanwhile, it added, the city will raise down payments for second-home buyers.

## 住建部：住房信息联网将扩至500城市

全国政协委员、住房和城乡建设部副部长齐骥近日表示，40个城市的住房信息联网已完成，而按照此前住建部的计划，今年6月底将完成500个城市的住房信息联网工作。业内人士分析指出，一旦全国范围住房信息联网之后，房产税的全面铺开的时间表将更明晰了，届时过渡时期的限购政策也将退出历史舞台。

### MOHURD: INDIVIDUAL HOME OWNERSHIP INFORMATION DATABASE TO COVER 500 CITIES

Mr. Qi Ji, the vice minister of MOHURD recently said that, a property information network of 40 major cities has already been built. According to MOHURD's plan, it will expand the scope of the database to include 500 cities by the end of June 2013. At that time, the property taxes on residential housing will come into effect in the country. And the property-purchasing limitations will be removed.

## 北京率先提出新建项目执行绿色建筑标准

据北京市规划委报告，北京市政府研究制定了《关于全面发展绿色建筑推动生态城市建设的意见》，要求从6月1日起，北京新建建筑基本达到绿色建筑一星级及以上，并在全国率先将绿色生态指标纳入土地招拍挂出让。《意见》中，北京率先提出把绿色建筑标准落实到新建项目中，并编制和实施绿色生态规划等要求。

### ALL NEW BUILDINGS IN BEIJING MUST MEET THE GREEN BUILDING STANDARD

According to Beijing Municipal Planning Commission, the government formulated Guidance on Promoting Eco-city Construction by Developing Green Building, saying all new buildings in Beijing must carry out green building standard from June 1. In the meantime, green ecological index was included to the range of land bidding, auction and listing-for-sale. It was the first time in China to propose green building standard in new projects. Green ecological plan is also required in design and construction stage.

# INFORMATION 资讯/设计

## 西班牙"北墙房"

"北墙房"由nodo17建筑事务所设计,位于西班牙马德里。这座房子是沿着北墙修建的,北墙上有很多basatls石块,常年满布青苔,很潮很滑,要想攀登上去,需要一身好技艺。这个方案目前只展示了这个房屋的一部分——北墙。建筑师想要建出一面特别自然、特别原始的墙壁,就好像是从悬崖处取下来一块断壁,摆放在了这里一样。

### NORTH FACE
The "NORTH FACE" is a house protected by a climbing wall: wrapped and always wet, where the moss grows on the basatls stones. You will need a "good hand" for climbing it. The house is devoloped along the "north face" wall. The model shows only a part of the house: the North Face wall. The main intention is to construct the most natural wall as posible. Almost like we cut a natural cliff and we carry it to the land.

## 三面有窗的小屋

这是由日本SNARK and OUVI建筑事务所合作建造的keyaki小屋,位于honiyo-shi小镇一个拐角处。这个小镇离东京只有一个半小时的车程。这里与北美很多城市一样,因为没有公共交通工具,人们都要开私家车出行。这座小屋有两个平行的车库,还有一个小花园。除了北边,房屋三侧都有细长的窗户,这样小屋内整天都能被太阳照到。

### House in Keyaki
The small town of Honiyo-Shi is about an hour and a half by car from downtown Tokyo, a place that resembles many North American cities so that all of its citizens rely on private automobiles due to the lack of public transit. Set back within the narrow corner site of the small town of Honiyo-Shi, the 'House in Keyaki' by Japanese studios SNARK and OUVI contains a parallel outdoor parking spot surrounded by a small garden that occupies the remaining plot. Three vertical strip windows located on the east, west, and south facades guarantee natural illumination throughout the interior at all times of day.

## 中国铜钱式房屋

这是由建筑师Juan Carlos Menacho Durán设计的中国铜钱式房屋,位于玻利维亚圣克鲁兹。古代中国普遍辩证思想的最高表现是天圆地方,相辅相成,而人居于天地之间。该房四四方方的外形可以给人以安全感和平衡感;放置于圆形的地基中,与天空融为一体,激发人们的创意。天圆地方的理念通过中国铜钱的方式在该建筑中得到体现。

### Chinese Coin House
Juan Carlos Menacho Durán has designed the Chinese Coin House in Santa Cruz, Bolivia. The sky is round and the earth is square. The supreme manifestation of the universal dialectic is the sky and earth, paired as a couple. In between both, lies man. This house inspires security and balance through its square form. Immersed in a round base, it inspires creativity and harmony from the sky. These two shapes –sky and earth– are present in the Chinese coin.

## 展馆式住宅

奥地利格拉茨Pernthaler建筑事务所打造的展馆式住宅,为生活在市中心的市民做出了很大的贡献。该项目发展成为当地商品交易区重建计划的一部分,旨在很好的平衡各种功能的需求。项目包括不同类型的租赁住宅、私人性质公寓、学生公寓、老年住宅,其间还充盈着办公区、一所幼儿园、服务区和美食城。

### Messequartier Housing Project
'The Messequartier' by Graz-based Pernthaler Architecture Studio is a high-quality contribution to living in urban centers. The project was developed as part of the re-structuring plan for the local trade fair district, whose attraction gradually lost substance within the past years. The concept of the residential complex aims at a decidedly well-balanced mix of functions. Various typologies of rental apartments, condominiums for private property, student residences and assisted living for elderly are complemented by office-areas, a kindergarten and space for service industry and gastronomy.

## 火山岩状的住宅

这座新建筑,位于韩国经济高速发展的度假胜地济州岛,名为the mineral,外形酷似当地火山岩上的水晶簇。建筑的中央庭院一直延伸入海,庭院里有一个10m宽的矩形游泳池。该建筑有四个独立的入口,模块化的建筑可以合而为一,也可以分隔成几个更小一些的居住区。屋顶上有个小花园,坐在上面可以夜观天象,或者远眺俯瞰周边森林。

### The Mineral Residence on Jeju Island
the newest addition to planning Korea's airest city-Berjaya Jeju Resort on the rapidly developing Jeju Island is 'the mineral', a residence/event venue that echoes the natural shape of the crystal clusters that form on the local volcanic rock. A central courtyard with a 10-meter wide rectangular pool that extends out towards the sea. Four separate entrances and the modular-like construction allows the complex to be joined into a single volume or subdivided into smaller residential units. A rooftop garden vertically extends the ground-scape creating an ideal environment from which to view the night sky or the forest canopy.

## 八岳山别墅

该别墅由日本当地MDS建筑事务所设计。这座木建筑要在不使用空调或其它人工取暖设施的前提下,适应四季气候变化。因此,建筑师设计了三个相邻但高度不同的外罩,它们有很深的悬挑,可在必要的时候控制进光量。扇形的体量布局留出一个朝南的平台,可以吸收冬季低矮的阳光。当南北立面两端的窗户都被打开,夏季凉爽的微风便能穿透整个住宅。

### Yatsugatake Villa
The 'Yatsugatake Villa' is designed by local practice MDS.The wooden structure needs to adapt to each season without the use of air conditioning or artificial heaters, with the ability to completely open itself to the elements in the warm months and close off – while still allowing the presence of sunlight – in the winter. Three adjacent shells of different heights developed as a result, with deep overhangs that control direct solar gain when not desired. the fan-like arrangement of the volumes exposes a greater surface area to the south to absorb the low winter light. With the majority of the windows on the southern and northern elevation, opening both ends invites a cool breeze through the entire structure.

## 珀斯场馆

澳大利亚珀斯当地的ARM建筑事务所与CCN金融集团合作建造了这座"珀斯场馆"。这个场馆计划用作网球场，也可以用作音乐会大厅，或者是其它一些小型活动，共有15 000个席位。棋盘格式的三角形外观赋予这张概念图以自由奔放的气息，多边形的设计风格打破了传统的对称型运动场所设计。该建筑屋顶可以在七分钟之内打开，也可以快速闭合。

### Perth Arena

The 15,000-seat 'Perth Arena' is designed by local studio ARM architects in collaboration with CCN. The arena, prepped for use as a tennis court, can also be transformed for use as a concert hall or for smaller more intimate events. Its tessellated triangular facade affords a freedom in the massing, creating highly geometric forms that read completely differently from each angle of approach, breaking from the traditionally symmetrical sporting venues. An operable roof can open in seven minutes and floods the stadium with natural light or closes during unfavorable weather conditions.

## 维也纳财经大学图书馆及学习中心

该项目由扎哈·哈迪德设计，目前仍在建设中。这个略微倾斜的立方体建筑外部是直线型构造，内部被分隔成很多小块，为多弧形构造。建筑外表将采用奥地利材料专家Rieder所研制的fibreC材料，这是一种加入玻璃纤维的混凝土面板，十分坚固。原材料很细很轻，但可防火，还具有可塑性和伸展性。

### Library and Learning Center of University of Economics and Business in Vienna

Zaha Hadid's design for University of Vienna campus is underway. The slanted cubic structure is comprised of a rectilinear exterior that gives way to filleted and curved interiors. the building is well into the construction phase, with parts of the fibreC facade by Austrian material specialist Rieder already apparent. FibreC is a concrete panel reinforced by glass fibers, combining the qualities of both materials in one tectonic; while the raw mineral mix is moldable, it is also thin, fireproof and lightweight while keeping its tensile strength.

## 个性木制展馆

Wolf建筑事务所建造的这座展馆，名为Perathoner，由木材雕琢而成。这个建筑展示了Gröden当地精巧的工艺制品及木质材料的悠久历史。建筑造型独特，内外都给人留下了深刻的印象：展馆内是价值不菲的艺术品，展馆外的个性木雕则引得路人纷纷驻足观望。建筑师先创建一个单一的立面作为这个中心，然后整合出四五个立面来。该项目已成为当地工艺制品的一部分。

### Perathoner

The estate in exceptional location (roundabout Pontives) is the obvious place for the new construction of the wood carving Perathoner. The new building should point out the long tradition of Gröden's craft industry and the material wood. The corporate identity of wood carving should be expressed by the outside as well as inside: in this building valuable art objects are produced. The facade also appears like that: a movement, a bend, a fold, a change. The building arouses curiosity because of the folds, dents, cuts and openings. It was a central aspect to create just one single facade at this traffic junction, to transform the 4 or 5 facades into one. The project becomes part of the craft industry.

## 西班牙演出中心

这是由Dra建筑事务所在西班牙Valdemeca修建的一座演出中心，混凝土一体墙上有彩色的砂砾石，还配有木制长条，与周边环境融为一体，是自然景色中一处形状不规则的建筑。建筑师Diaz Romero把自然引进建筑之中，在每一面墙壁上都开了大大的窗户，是典型的当地建筑风格。

### Interpretation Center

Breaking the geometry of the place deliberately, some forms monolithic concrete facts, formwork wood duplex with wooden slats and colored tinted sandstone, looks the landscape, integrating the irregular geometry of nature. In the proyect, the landscape becomes the generating element of the architecture, and the memory of the place will help us contemporizing the typical houses of the locality. All the rooms of the interpretation halls have large windows that look to the place.

## 园艺博览会画廊

依河而建的科布伦茨最近举办了德国的联邦园艺博览会。比利时Dethier建筑事务所赢得了开放式画廊的设计权，所造建筑成为了这个盛会的重要组成部分。这个开放式画廊是个三角形的观景台，建于高地，能够俯瞰令人感叹的城市全景和绿色植被。悬挑起来的观景台向山谷方向延伸了15 m，离地面有10 m，整体看去像是一个回形走廊。

### Belvedere for Koblenz Bundesgartenschau

the riverside city of Koblenz was host to the recent Bundesgartenschau. Belgium-based Dethier architectures winning entry for a open-sided gallery was built as part of the festival. The triangular belvedere is positioned on a plateau, optimal for stunning views of the city and greenery. The cantilevered form creates orthogonal elevated pathways, extending 15 meters out over the valley, and rises 10 meters above the ground.

## Luftfartstilsynet挪威民航总部

NAV建筑位于海滨，内有隧道，作为可控接待桥，联通至Luftfartstilsynet的独立建筑。现代化的结构使Luftfartstilsynet成为一座既与其他建筑相连又自由独立的建筑。从入口隧道到楼内开放高台都彰显了文化和多样性的理念。建筑整体如经过外科手术般地切过，像颗钻石，带来北面富有活力的光线。Luftfartstilsynet总部代表了对SPEC办公楼的新定义：富有雄心和高环境标准（LEEDS标准）。

### Luftfartstilsynet Headquarters

NAV building is a compact waterfront property, with a tunnel that becomes a programmed (reception) bridge, connecting to the freestanding Luftfartstilsynet. Urbanistically the building completes the urban fabric whilst providing Luftfartstilsynet with a stand-alone building - connected and free. The strategy, committed to CULTURE and diversity, extends the entry tunnel to a raised, open level in the building. Triggered by the sight lines of the city, the building is surgically cut, like a diamond, capturing the dynamic light of the north in surprising ways. Luftfartstilsynet Headquarters represents a new definition for the SPEC office building: high design ambition and high environmental standard (LEEDS standard).

# UPSCALE COMMERCIAL SPACE WITH SIMPLE AND CLASSICAL STYLE

| Greenland Jiashang International, Shanghai

## 体现简约古典风格的高端商业空间
—— 上海绿地嘉尚国际

| | |
|---|---|
| 项目地点：中国上海市 | Location: Shanghai, China |
| 开 发 商：绿地集团 | Developer: Greenland Group |
| 建筑设计：水石国际 | Architectural Design: W&R Group |
| 项目规模：60 000 m² | Area: 60,000 m² |

**项目概况**

本案为商业建筑，主塔楼为酒店式办公，附带商业裙房和地下车库，基地位于嘉定南门地块，踞守嘉定CBD，区政府、海关大楼、嘉定信息大楼等政要机关环伺，高端写字楼陆续崛起，庄重严谨的行政氛围日渐浓厚。同样由绿地集团打造的嘉创国际作为嘉领国际的姊妹楼，两者在沪宜公路与叶城路路口隔街对望，扼守嘉定南门，形成双子塔的高端形象，加上周边便利的交通条件以及得天独厚的自然景观，后期的标志性地位不言而喻。

**建筑设计**

本案在建筑设计上采用了简约古典风，既彰显嘉定这一文化古镇的悠久历史感，又体现建筑自身的高贵品质感，并延续嘉创国际含蓄内敛的主旋律。总体布局顺沿了南部块状商业的分布形态，采用了开放式商业内街方式，在尊重地块周边原有商业风貌的同时，也营造出一种亲切、活泼的商业氛围。建筑裙房部分以灵活多变的设计语言，丰富时尚的形态变化，营造出一步一景、步移景异的商业空间。

在商业裙房的外立面处理上，运用虚实结合的设计手法，通过大面积石材所体现的"实"和部分玻璃幕墙所体现的"虚"相互穿插，使之形成强烈的对比，并具有浑厚的雕塑感，彰显大气婉约的内敛气质和高端商业的尊贵品质。本地块的商业设计在内庭中营造出广场、回廊、中庭、院落等多种聚落模式，使每一间商铺都有充分的延展面和引人入胜的趣味空间。塔楼部分采用artdeco的竖向线条，保持主建筑庄重典雅的风格，并提升延展向上的感觉，形成的独特韵味对打造城市名片具有良好的推动作用。

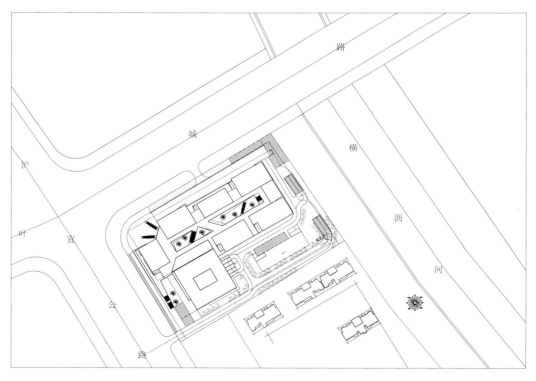

总平面图 Site Plan

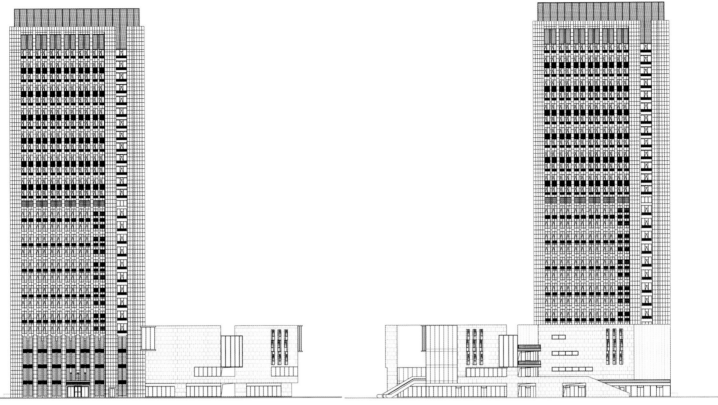

立面图 1 Elevation 1

立面图 2 Elevation 2

### Overview

As commercial architectures, the project is composed of hotel-style offices building---the main tower building, and commercial podium, the garage underground. It is located in Jiading Nanmen, entrenched in Jiading CBD and circled by District Government, Customs House, Jiading Information Building etc. The zone is featured for its increasing solemn political atmosphere and rising-up upscale office buildings. Jiachuang International developed by Greenland and Jialing International are facing each other across the intersection between Huyi Road and Yecheng Road. They are standing in Jiading Nanmen as an image of upscale twin towers. With a convenient traffic environment and superior nature landscape, it is self-evident to be a landmark in the future.

### Architectural Design

The design adopts the style of simplicity and classism to manifest the long history of Jiading as a cultural ancient town. In addition it shows the nobility and high-quality of the architecture and carries on the reserved main theme of Jaichuang International. The master layout follows the commercial layout of the southern plot with an open internal-commercial-streets arrangement to create a friendly and vivid commercial atmosphere, in the meanwhile pays respect to the existing commercial style surrounding. The podium presents a commercial space that scenes are changed step by step.

立面图 3  Elevation 3

立面图 4  Elevation 4

# MASTER AND MASTERPIECE | 名家名盘

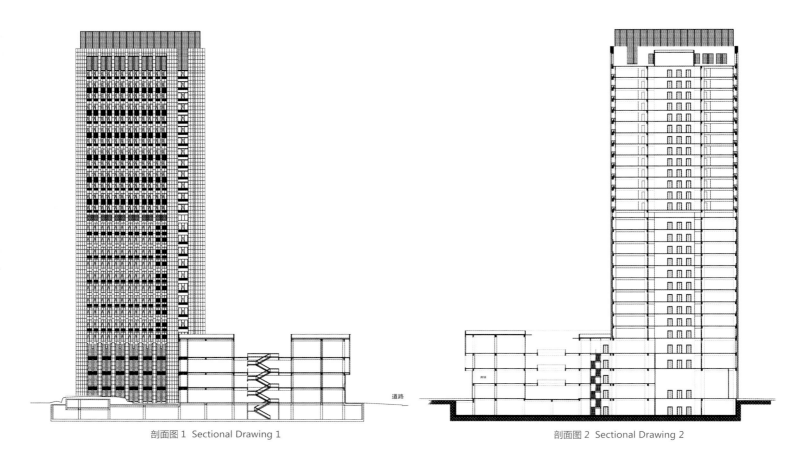

剖面图 1  Sectional Drawing 1

剖面图 2  Sectional Drawing 2

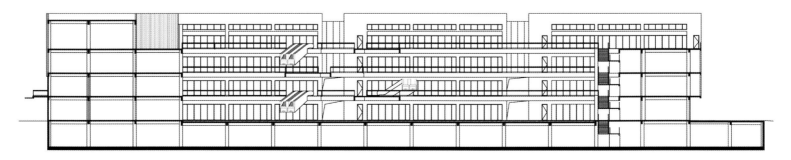

剖面图 3 Sectional Drawing 3

MASTER AND MASTERPIECE | 名家名盘

The facade design of the commercial podium is an integration of reality and unreality, through the intersection between the great-area stones and the partial glass curtain wall. It creates a strong contradiction and the deep sense of sculpture, manifesting the quality as graceful, reserved and noble of upscale commerce. The commercial design of the inner space in the plot adopts various modes of settlement like plaza, corridor, atrium and courtyard etc to make every shop interesting and fascinating with enough extensive surface. Part of the tower building adopts the artdeco vertical lines to maintain the solemn and elegant style of the main architecture, and generates the sense of upward extension. Its unique charming has played a favorable role in building the city card.

# COMMUNITY IN PRAIRIE AND PASTORAL STYLE WITH FLEXIBLE DEGREE | Nantong Shanshui No. 1

## 张弛有度的草原式田园风格社区
—— 南通山水一号

项目地点：中国江苏省南通市
建筑设计：上海中房建筑设计有限公司
总建筑面积：164 934 m²

Location: Nantong, Jiangsu, China
Architectural Design: Shanghai ZF Architectural Design Co., Ltd.
Total Floor Area: 164,934 m²

**项目概况**

项目位于南通五山风景区板块，基地北临星湖大道，东接长青路，西侧南侧为规划路。地块中部有南北向10m宽自然河道穿过，景观资源十分优越。项目坚持以人为本，强调人与环境和谐，力争做到总体规划布置合理，交通流畅清晰，功能分区明确，适应市场发展，为用户提供多种精心设计的户型。

**规划布局**

规划设计从城市空间入手，考虑城市界面的积极因素。由于北侧星湖大道较宽，且建筑退界30m绿化景观带，为了保持沿街尺度关系的协调感，沿星湖大道建筑均匀布局了四幢8层住宅。而基地东、西、南三侧

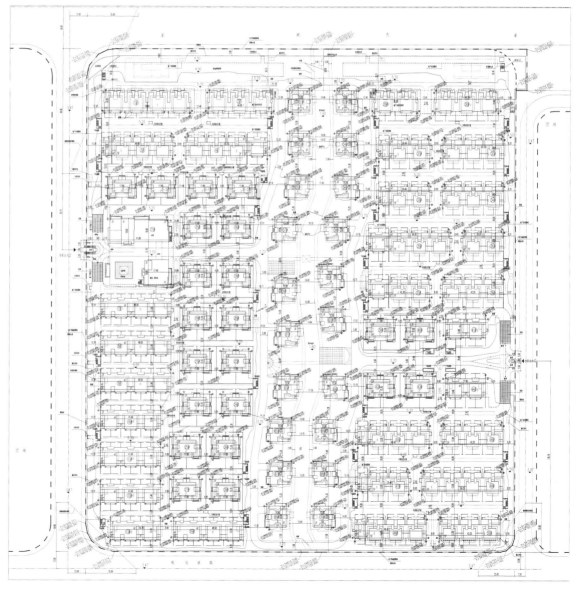

总平面图 Site Plan

# MASTER AND MASTERPIECE | 名家名盘

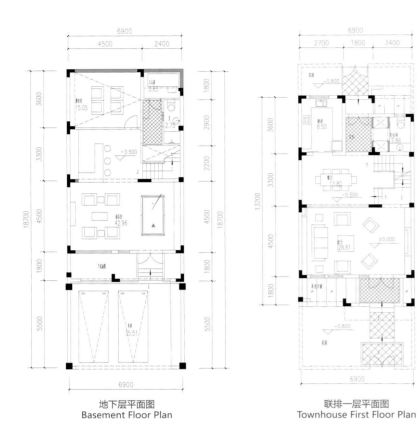

地下层平面图
Basement Floor Plan

联排一层平面图
Townhouse First Floor Plan

道路为城市次要道路，设计采用局部点板结合的排布方式。通过景观绿化设计突出居住组团，形成较丰富的沿街界面和内部空间；设计着重在西侧主出入口处设置了入口广场与主题VIP场所，强化入口空间，打破沿街界面的连续感，起到了给人眼睛一亮的标识作用；而东侧出入口设计了景观林荫大道及标志性塔楼，突显了高档住宅社区的品质特征。

同时，规划设计考虑如何解决小区内部组团结构与产品分布特点的结合。设计将面积较大的独幢花园住宅临河布置，以取得较好的景观面，同时点式住宅由于南北向间距较大，景观视线通透，其两侧的住宅组仍可看到河景，并结合景观细化设计，营造出小桥、流水、人家的居住意境。

**建筑设计**

住宅建筑立面采用草原式田园风格，以体现高级住宅的品质特征。通过形体的尺度推敲，多种材质的配合运用，形成丰富而精致的质感，住宅建筑均采用坡顶，并通过高低错落的建筑布局形成较活泼的城市第五立面，与整个风景区相协调。公建立

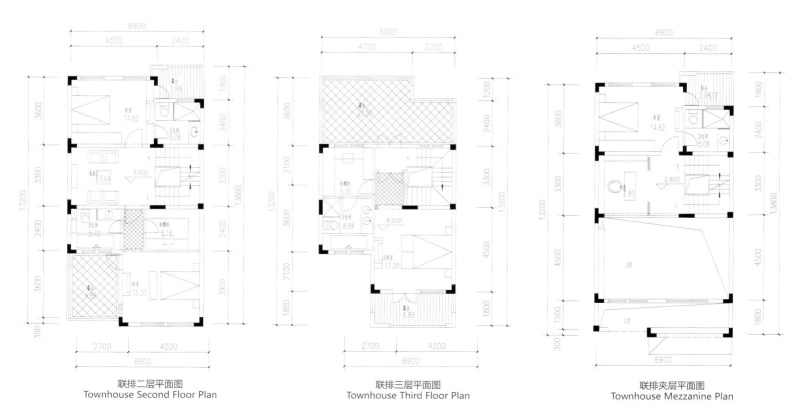

联排二层平面图
Townhouse Second Floor Plan

联排三层平面图
Townhouse Third Floor Plan

联排夹层平面图
Townhouse Mezzanine Plan

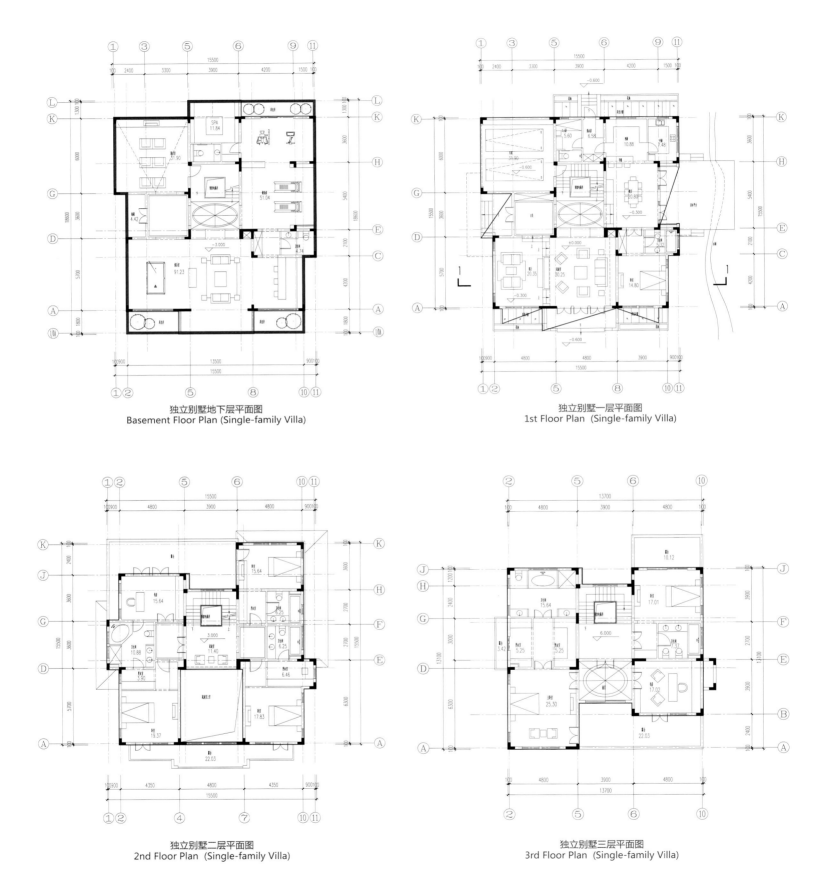

面设计现代简洁，烘托出城市感十足的社区氛围，并通过材质的选择与细节的推敲与住宅立面相融合。建筑造型与立面设计上力求在满足功能的前提下，丰富创新，张驰有度，使其既符合区域整体规划，又具有自身鲜明的建筑个性和社区品味。

**景观设计**

作为整个小区的景观核心，会所设置于小区西侧主出入口，结合入口广场、景观水池、并利用场地的高低差造景，形成整个项目的形象中心。同时沿星湖大道30 m景观带的设计以自然坡地为主，结合运动主题的加入，使景观由静而动，富有活力。整个居住街坊的空间形态、城市界面讲求收入有致，组团空间注重变化错落，并积极利用现状水系和景观绿带，结合标志性入口空间的深化设计，提升整体住区的综合品质。

会所一层平面图
Plan of the 1st Floor of Club

MASTER AND MASTERPIECE | 名家名盘

独立别墅正立面图 1
Front Elevation 1 (Single-family Villa)

独立别墅正立面图 2
Front Elevation 2 (Single-family Villa)

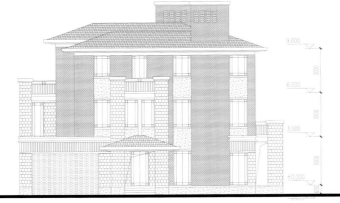

独立别墅侧立面图 1
Side Elevtion 1 (Single-family Villa)

独立别墅侧立面图 2
Side Elevtion 2 (Single-family Villa)

## Overview

The project is located in Wushan Scenic Plot of Nantong, Xinghu Avenue to its north, Changqing Road to its east, and the planning roads to its south and west end., with superior landscape resources. A 10m wide natural river towards south and north cross through in the center of the plot. The project highlights human-oriented and harmony between human and environment. It aims to set a reasonable arrangement about the master plan with clear and smooth traffic, definite functional zoning, and offering various house types with exquisite design responding to the market development.

## Planning

The planning starts from the urban space, considering the positive factors of the urban interface. Given that Xinghu Avenue to its north is wide and the buildings setbacks 30m for greenbelt, four 8-floor buildings are placed even along Xinghu Avenue to hold the harmony of scale relation along the street. The south, west and east side of the plot is secondary main roads of the city with the design as a combination of partial point and plate. Residential groups are highlighted through the landscape and greening design, forming rich street interfaces and internal space. The design sets the entrance plaza and themed VIP spaces in the main entrance of the west side acting as an eye-catching mark to strengthen the entrance space and break the continuity sense of the street interfaces. The Landscape Boulevard and iconic tower in the gateway of east side have underlined the quality features of upscale residential community.

Meanwhile, the planning considers how to solve the combination between the internal group structure and the product distribution characters. Detached harden houses with larger area are located along the river to obtain better landscape; point block apartments with large distances from each other towards south and north lead to transparent landscape views and river scenery. Integrated with landscape detail design, they create an artistic conception of bridge, stream and household for dwelling.

## Architectural Design

The building facades adopt the prairie and pastoral style to manifest the quality features of upscale residences. Rich and delicate sense of texture is formed through the scale elaboration of the body, and the coordination among multiple materials. The residential buildings with slope top build the vivid fifth facade of the city through the scattered layout, in harmony with the entire scenic spot. The facade of public buildings is designed in modern and

MASTER AND MASTERPIECE | **名家名盘**

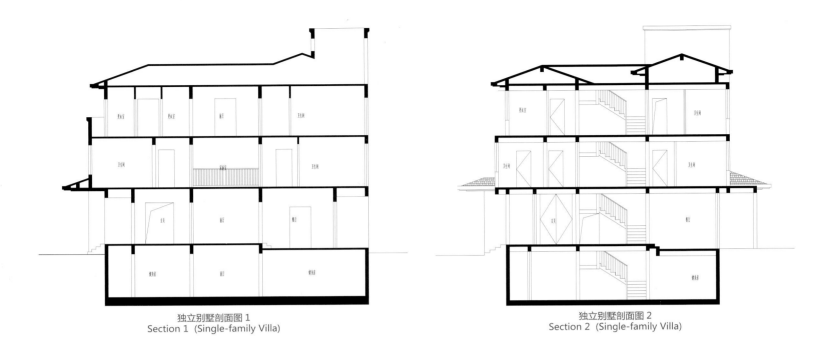

独立别墅剖面图 1
Section 1 (Single-family Villa)

独立别墅剖面图 2
Section 2 (Single-family Villa)

simple style to set off the community atmosphere of abundant urban sense, and to integrate with the residential facade through the selection of materials and the elaboration on detail. More than fulfilling the functions, the building shapes and facade design are innovative and flexible to conform to the master plan of the district as well as to be distinguished for its distinct architectural character and community taste.

**Landscape Design**

As the landscape core of the general community, clubs are located in the west main gateway, acting as the image center through landscaping in different heights with the entrance plaza, and landscape pool. 30m landscape belt along Xinghu Avenue is designed mainly based on natural sloping fields, with sports theme to add vitality to landscape. The space forms of the entire dwelling neighborhood and urban interface requires order and the group spaces focus on scatter and changes; the design utilizes the existing water system and landscape belt along with the further design of iconic entrance space to promote the comprehensive quality of the entire dwelling community.

INTERVIEW | 专访

# 商业综合体的设计要做好业态整合与因地制宜
——访HMA董事、设计总监 东英树

■ 人物简介

东英树 HMA董事、设计总监
1996 近畿大学理工学部建筑系毕业
1996 若林弘幸建筑研究所就职
1998 Plantec综合规划事务所
2002 取得"日本一级注册建筑师"资质
2003 日本HMA建筑设计事务所合伙人创始人，设计总监
2010 上海世博会城市最佳实践区B4馆，荣获日本GOODDESIGN奖
2011 UIA第24回世界建筑会议东京大会 演讲

■ 代表作品

上海建国中路8号桥
上海世博EXPO城市最佳实践区中部系列B4展馆
SKYWAYHOTEL
天津水游城
金地绍兴柯桥商业
徐州和信广场
宜昌水悦城
好世马陆站北地块

《新楼盘》：一个成功的建筑总有很多值得回味和探究的地方，您作为一位知名的设计师，曾经去过很多城市，也见到过形形色色的建筑，那么您对于建筑的理解是怎样的呢？一个好的建筑设计的评判标准又是什么呢？

东英树：我认为优秀的建筑必须能够给予来访者美妙的体验与感动。

在近20年的时间里，我造访过很多国家，也见到过很多的建筑。给我的感受是，无论经过多少岁月的洗礼，优秀的建筑仍然能够给人以美妙的体验和感动。在不断的建造和拆毁的过程中，能够让人们一直喜爱并得到保存的建筑才称得上是最好的建筑。

当然即使是最好的建筑也会由于各种各样的原因而面临被毁坏的情况。如果一个建筑被拆除，而市民们都倍感可惜，我觉得这也是好的建筑。

《新楼盘》：近年来，以商业综合体为代表的商业建筑开发非常火爆，你们是怎么样看待这种现象的？未来商业综合建筑的前景又是怎样的？

东英树：直到5年前，中国开发的最多的还是住宅，之后由于库存增大、国家宏观调控，住宅开发比例开始降温。其后商业设施的开发建设增速，甚至某些城市有商业设施已经供过于求的说法。究其原因，这是因为随着收入提高，人们在住房的基本需求被满足后开始转向消费需求。其次，为了满足人们精神上的需求，用来消磨时间的度假设施和休闲场所会越来越多，这也是以商业综合体为代表的商业建筑异常火爆的原因之一。

我认为未来的商业综合设施不仅需要满足人们对物质的需求，更要满足人们对精神的需求，需要时间消费型的商业设施。

虽然现在的时代已经是可以通过网络来实现人们购物与交流，但是我认为还是需要为人们提供能够聚集在一起消磨时间的空间。

《新楼盘》：在进行一个大的综合项目以前，一般都要进行大量设计前的准备工作，就商业综合体而言，其前期的策划对于项目后期的建筑设计以及建成后的运营有何重要的意义？

东英树：前期的策划决定了项目开发的大方向，可谓是开发过程中最重要的环节。

一般来讲，我们在进行商业建筑设计的时

天津水游城

天津水游城

天津水游城

天津水游城

候，都会进行大量的前期准备和策划工作。特别是其中的市场调查环节，其在决定目标客户上是最重要的，完全可以说这是决定项目能否成功的关键，同时也直接影响到后期的运营和管理等环节。目标客户的定位等前期任务如果能够很好完成的话，设计都可以按照前期策划来展开，直到最终的完成。同时，建成后的运营也不会偏离规划，这对于整个项目而言具有非常重要的作用。

**《新楼盘》：商业综合体与一般的公共建筑相比，在设计内容和形式上有哪些不同？**

东英树：与其他公共建筑相比，就像综合体这个名称所表明的那样，不同之处在于其相关的运营方众多，在于其综合的性质。

在规划阶段，需要将酒店、办公、商业、住宅等不同功能和形态的运营方召集在一起不断地沟通、协商和整合。由于不同的运营方的要求会相互矛盾，所以我认为在商业综合体的设计中最重要的部分就是如何将这些不同的需求整合成一个整体。

**《新楼盘》：在空间结构以及形态的设计上，商业综合体应该怎样去与周围的环境协调一致？**

东英树：商业综合体的庞大体量在城市可能会显得突兀，要将其与周围的环境完全协调一致是十分困难的。

但即使不能与周围的环境完全协调一致，我想能够创造城市的新形象也是很重要的。

我在设计之前会深入理解当地特有的文化和历史，并将其在设计中反映出来，我希望我的设计能成为这个街道的符号。

**《新楼盘》：作为一名外籍的设计师，您觉得中国建筑在设计过程方面与国外是否有不同之处？这种不同体现在哪些方面？**

东英树：这种设计过程的不同主要是中国政府管理部门人员过多干涉建筑设计，较多的行政手段出现在一些大型项目的设计之中。而在国外，只要遵守相关法律与职业操守去进行设计就行了，除了业主，其他的人是不会对建筑提意见的。在中国，由于土地是政府的，政府就会给建筑提较多的意见。作为一名外籍的设计师，我觉得外国建筑师需要理解这个巨大的差异，设计出满足业主、政府和自己事务所等三方面要求的建筑。

**《新楼盘》：在2010年的上海世博会上，您负责设计了当时的世博会城市最佳实践区B4馆，并且获得了设计大奖，能否回忆一下这个项目当时设计上的一些细节吗？您觉得这个项目获奖的关键是什么？**

东英树：上海世博会上，当时设计的项目是以环保作为主题来改造陈旧的厂房。由于环境保护的问题，是禁止在城市里使用红砖的，这也是这个项目当时面临的一个问题。于是，我们对于已经存在的红砖进行再利用，做出了一个非常有特色的外立面，很好的契合了环保的主题。这个项目能够获奖主要是充分发挥中国丰富的人力资源，重砌了需要大量人工的红砖墙，成为日本设计者的创意与中国工匠的手艺相互融合的作品，因此在日本和欧洲也受到了好评。

**《新楼盘》：最后有没有新的设计项目与大家分享呢？可否列举1～2个，简要谈谈其设计方面的特色？**

东英树：近期主要有两个项目，分别是天津水游城和盘锦水游城，它们是继南京水游城之后的标志性商业综合体。

2011年秋竣工的天津水游城，该项目沿城市主干道展开面将近500 m。为了不让消费者在步行购物的同时感到疲倦乏味，将建筑形态在视觉上分为3段，在各段的功能节点处设计具有象征性的建筑构造物，将顾客从设施的起始端部顺畅地引导向另一端部。

同时引入多种设计手法，利用浓重的商业氛围、有序的商业环境吸引和留住消费者。街角广场、入口广场以及各个中庭空间交替出现，营造高档的商业氛围。立面上利用现代设计手法，精练的设计造型表现大气新颖的建筑体态。同时采用时尚的建筑材料、精巧的建筑形态使之与周边的建筑群相呼应。

对于此类商业综合体，我们都旨在打造以光、水、风为主题的时间消费型商业设施。同时希望外观是每隔100 m就有一个标志性建筑物，并能够提高人的流动性。

# The Design of Commercial Complex Should Pay Attention to the Industry Integration and Act According to the Local Conditions
—— Interview with Hideki Azuma, Board Director and Design Director of HMA

## Profile

Hideki Azuma, Board Director and Design Director of HMA
1996 Graduated from Department of Architecture, Faculty of Science and Engineering, Kinki University
1996 Joined in the Hiroyuki Wakabayashi Architects
1998 Plantec Archietcs
2002 Got the "Japan Registered Architect Qualification"
2003 Founder and Design Director of Japanese HMA Architects & Designers
2010 Pavilion B4 in the Best City Practice Area of Shanghai World Expo won Japan GOODDESIGN Prize
2011 Had a Speech in the twenty-fourth UIA World Architecture Conference in Tokyo

## Selected Works

Shanghai Jianguo Road No. 8 Bridge
2010 Shanghai Expo the best city practice area pavilion B4
SKYWAYHOTEL
Tianjin Aquatic City
Jindi Shaoxing keqiao Commercial street
Xuzhou Hexin Commercial Plaza
Yichang Aqua City
Northern Architectural Block of HOUSE in Malu Station

---

**New House:** A successful architectural project is always worth to be recalled and explored. As a well-known designer, you have been to many cities and seen all kinds of architectures, so how do you define the word of architecture? What are the criteria of a good architectural design?

**Hideki Azuma:** I think good architecture must be able to give visitors wonderful experience and move them.
In the past 20 years, I have visited many countries and also seen a lot of buildings. I feel that, regardless of how many years of baptism, the outstanding architecture can still give people a wonderful experience and move them. During the construction and demolition of the buildings, only the buildings that people love and be preserved along should be regarded as the best architecture.
Of course, even the best buildings, they will face being destroyed due to various reasons. If a building is demolished in the people's voice of pity, I think this is also a good building.

**New House:** In recent years, the commercial complex is very hot as the representative of the commercial building development, how do you think of this phenomenon? And what the future commercial building prospect will be?

**Hideki Azuma:** Until 5 years ago, China mainly developed house projects, later, because of the inventory increase and the national macro-control, the proportion of residential development began to go down. Then, the commercial facilities development and construction speeded up, and even some city commercial facilities have piled up in excess of people's requirement. Firstly, because people's income increases and their basic housing requirements have been satisfied, then they surely will turn to the consumption demand. Secondly, in order to meet the spiritual needs, resort facilities and leisure places for killing time will become more and more; this is one of the reasons why the commercial complex become hot as the representative of the commercial building.
I think the future commercial complex needs not only to meet people's substantial requirements, but also meet people's spiritual demand. People need the type of commercial facilities for time killing.
Although we are in the age that people's shopping and communication can be achieved through the network, I think it is still necessary to provide a space to gather people for spending time together.

**New House:** Before starting a comprehensive project, designers generally have a lot of design preparation, commercial complex, its early planning have important significance for the design and operation of projects completed?

**Hideki Azuma:** The early design planning determines the direction of project development, which is the most important link in the process of development. Generally speaking, in the design of commercial buildings, we will do a lot of preparation and planning work ahead. Especially the market survey is the most important in determining the target customers, which is critical for the success of the project, and also directly affect the following operation and management. If the position of the target customer and other preparatory tasks can be well done, the design will go on in accordance with the preliminary planning until the final completion. Besides, the operation will not deviate from the plan after the completion, which plays a very important role for the whole project.

*New House:* what is different in content and form between the commercial complex and the general public buildings?

**Hideki Azuma:** Compared with other public buildings, just as the word "complex" indicates, it is different because its relative operators are numerous and it develops in a comprehensive way.

In the planning stage, operators of different functions and forms from hotel, office, commercial and residential industries constantly gather together to communicate, negotiate and integrate with each other. Because of the conflict from different operating requirements, I think the most important in the design of commercial complex is how to integrate these different needs into a whole.

*New House:* In the aspects of space structure and shape design, how does the commercial complex interact and coordinate with the surrounding environment?

**Hideki Azuma:** The large scale of the commercial complex may appear to be abrupt in the city, and how to coordinate with the surrounding environment is very difficult.

But even if it can not coordinate with the surrounding environment entirely, I think it is still very important be able to create a new image for the city.

I will learn the unique local culture and history before the design, and reflect these elements in the design; I hope my design can become the sign of the street.

*New House:* As a foreign designer, what do you think are the differences between China and the foreign countries in the architecture design process? And what aspect do these differences reflect in?

**Hideki Azuma:** The main difference in the design process is that the government personnel give over interference in the construction design; more administrative means are appear in the design of some large-scale projects. While in a foreign country, as long as it compliances with relevant laws and the occupation ethics in the design, other people will not comment on the architecture design except the owners. In China, because the land belongs to the government, the government will give more comments on the building design. As a foreign designer, I think foreign architects need to understand this huge difference, and try to meet the requirements of governments, their owners and themselves.

*New House:* In the 2010 Shanghai World Expo, you were responsible for the design of Hall B4 in the best city practice area of Shanghai World Expo and won a design award. Can you recall some details in this project design? What do you think is the key point to win award for this project?

**Hideki Azuma:** For the Shanghai World Expo, the design of the project is to take environmental protection as the theme to transform an old factory. Due to the problem of environmental protection, the use of red brick was the prohibited in the city, which is a problem the project faced at that time. So, we reused the existing red brick to make a unique facade, thus well fitting the theme of environmental protection. The reason why this project wins the award is mainly because it fully plays the advantages of the Chinese abundant human resources and rebuilds the red brick wall that requires a lot of artificial labors. This project is the joint work of the Japanese designer's creativity and Chinese artisans' craft skill; therefore, it receives a good reputation in Japan and Europe.

*New House:* Are there any more new design projects to share with us? Can you give 1 or 2 examples to briefly talk about their design features?

**Hideki Azuma:** Recently, there are two projects, the Tianjin Aqua City and Panjin Aqua City, which are another landmark commercial complexes after the Nanjing Aqua City.

Tianjin Aqua City was completed in the autumn of 2011 and its extended surface along the main city roads is nearly 500 m. In order not to avoid the consumers being tired and boring when they are shopping on foot, the project shape design is divided into 3 sections visually; it sets symbolic building structures in each functional design node, which can smoothly guide the customers from the starting point of the facilities d to the other end.

At the same time, it introduces a variety of design techniques, using strong commercial atmosphere and orderly business environment to attract and retain the consumers. The appearance of the corner square, entrance plaza and the atrium space in turn creates high-grade commercial atmosphere. The facade designed with modern design techniques and the concise design modeling form an elegant and novel architectural image. Besides, the fashionable building materials and exquisite architectural shape make the project echo with the surrounding buildings.

For this kind of commercial complex, we are aimed at creating time consuming business facilities with light, water and wind as the design theme. And we also hope that there will one landmark building every 100 m, and improve the mobility of the crowd.

2010 上海世博最佳实践区中部系列 B4 展馆

2010 上海世博最佳实践区中部系列 B4 展馆

# INTERVIEW | 专访

## 景观设计要注重差异化的空间营造
——访广州邦景园林绿化设计有限公司董事、设计总监 谢锐何

**■ 人物简介**

**谢锐何 邦景园林董事、设计总监**

谢锐何先生曾在多间著名的景观工程、设计机构工作，担任过主任设计师、项目总监等职位，现任邦景园林董事、设计总监。工作范围涉及不同尺度和各种项目类型的全流程，在景观设计与管理方面具有较强的把控、组织协调能力。能够胜任景观方案设计、景观深化设计、景观施工图设计、工程现场等项目的不同阶段的创作、细化、实施和管理工作。

谢锐何先生出色地完成了上百个项目的景观设计和项目管理工作，其中包括万科地产、侨鑫地产、合生创展、保利地产、时代地产、中信地产、江中集团、南山集团、鸿鑫集团等众多国内知名地产公司的项目。

**公司简介：**

广州邦景园林绿化设计有限公司是一家专门从事景观规划设计、工程施工、绿化养护、苗木生产的专业景观公司。其拥有一批实践经验丰富的境内外设计精英团队，同时配套有完善的专业设计队伍、施工队伍、综合的资源和真诚的服务。公司注重施工技术和工程管理方法，形成了一套以施工技术为保障、以项目管理为手段的工程管理系统，确保高效率、高品质完成建设项目。公司从创立之日起，努力走专业化、品牌化道路，以前瞻性的创作理念和差异化的造景手法，因地、因客户制宜，为客户营造诗意的景观环境，得到了广大客户的高度评价。

**《新楼盘》：您作为一名资深的景观设计师，同时身兼设计公司的高管职务，请您谈一下对景观设计的理解？**

谢锐何：很荣幸自己能成为一名景观营造工作者，同时也深深地肩负着这个职业所带来的压力与责任感。

景观设计，不是一个仅仅靠着平面、手绘等等的简单审美能够打动人的行业；不是一个靠着植物堆砌能够塑造空间的行业；也不是一个由着奢华材料可以打造出来的行业……

景观给予人类最直观的展现就是当人置身其中所感受到的风情、意境，因此景观（规划）设计就是在做各种尺度、各种价值的空间营造，以及如何运用硬质和软质元素合理、有效整合空间的一种行为。我想在这里强调：景观设计是一个交叉学科，也是一个漫长、复杂的修改完善过程。很多刚入行的新人怀揣着对设计的激情，但找不到方向，就很容易就被这一修改完善的过程折磨得失去耐心，急功近利，过分追求表面的美化、装饰，放弃了最初的理想，导致行业前景堪忧。

**《新楼盘》：您和您的团队设计了多种类别的景观项目，那么您觉得景观在住区以及公共环境方面分别承担着怎样的角色呢？**

谢锐何：邦景园林自创立以来，为改变不同尺度和类型的空间提供景观问题的解决方案、实施工作。

生态环境已然被高速城市化进程所更改，然而爱美之心人皆有之，随着人们生活质量、生活水平的不断提高，优越的生态景观环境成为人们的追求。不仅房地产开发企业在市场竞争中竞相打起了"景观牌"、"生态牌"，一些市政项目和公共空间也越来越注重环境景观的营造。十八大报告提出"美丽中国"的概念，强调山要绿起来，人要富起来。我认为，景观在住区以及公共环境方面的不同之处在于：团队致力于解决问题的对象不同，承担的责任不同。例如在住区景观设计时，虽然应注重生态角度的切入点，但更多是从满足项目功能性、艺术性出发；而在公共环境方面，景观设计承载着延续公共空间、公共秩序和生态格局等多种宏大之意义。我跟邦景园林的团队无论是在处理住区景观设计还是公共空间的景观设计时，都坚持运用差异化的空间手法，针对每个空间进行设计的具体解答。当然，这样做需要我们花费更多的时间和精力，这也是我们对场地负责，对客户负责，对邦景园林出品负责的态度。

**《新楼盘》：您一直在倡导差异化的空间设计手法，在很多项目里也有体现，请您简单介绍**

一下这种设计手法的特点以及如何突出自己的特色？

谢锐何：差异化的空间设计手法是邦景园林一直践行的设计理念。

所谓差异化，就是每一个设计师的个性追求，就是对不同类型客户的尊重，就是对每个空间独特的价值观，甚至是信仰的执着。简单的说，就是一种追求，对空间的追求，对自己的追求，对人文的追求。没有这种追求，就容易沦为一种纯功利性的设计，就容易沦为单调乏味的作品，甚至会留下许多对空间价值追求的遗憾。这也就是邦景园林坚持各专业、各阶段跨界工作的原因。正因为有各个专业的通力合作，才会有这些各式各样、魅力独特的空间。

**《新楼盘》：请问您所领军的邦景园林项目如何运作？邦景园林的人才观是怎样的？**

谢锐何：邦景园林的项目离不开邦景团队的密切配合。我们的项目一般都是由团队成员共同完成。不管项目的规模有多大或项目的地点在哪里，我们都会通过实地调研，与客户沟通，深入解读研究项目地块的特性，以及将来使用者的感受，努力找到项目的最佳解决方案，并赋予项目最和谐的空间品质。

邦景园林的每个作品都源于团队的集体智慧。这些作品都与邦景园林的每一个团队成员的个性修为、品味素养有紧密的关系。这个团队共同学习、共同进步、共同创造，不抛弃亦不放弃每一个团队成员，也正是如此，也才有紧密凝聚在一起的邦景园林。我们也欢迎有志之士加入我们，成为这个团体同创新、同实践的一员。

**《新楼盘》：请介绍一下邦景园林最近的景观作品，并谈谈其特色是什么？**

谢锐何：近期邦景园林建成的作品不少，我在此谈谈其中两个项目，一个是位于佛山南海仙湖旅游度假区的仙湖畔的和丰颖苑，另一个是位于佛山南海的天安数码城四期景观，两个景观各有特色，也恰好代表了邦景园林在不同类别的景观项目中对于"差异性空间"的探索与实践。

和丰颖苑项目景观采用"中西合璧，古韵新做"的处理手法，借鉴中国古典园林的空间营造方式，并以西方现代的景观语言加以诠释，古韵新做。以丰富的绚烂花木植物为基底，糅合中国传统园林文化，搭载西班牙式浪漫风情细节。并在花草林木间营造若干个特色多样化的公共景观庭院空间，为业主们提供丰富的景观视野和放松身心的休闲活动区域。

天安数码城四期景观在做的时候遇到的难题是：如何在一个已有三期建成区的科技园区做出差异化空间。在传统空间里传承新意容易，在一个已经成型多年的创意空间里寻求创意确实为不易。经过几次勘察、体悟场地和方案比稿，我们选用了最为简洁却直指人心的图形。低成本、易维护的造型植物坡构成了整个空间连续的元素。也正是这个简单的元素构成，结合点状的艺术节点空间，给予了在高楼大厦里办公的人们一个开放、自由的公共场所，用铿锵有力的空间价值诠释了新一期的景观，做出属于邦景园林的注解。

**《新楼盘》：目前，很多的设计公司都转向多样化、综合型的方向发展，谈谈您对这一现象的看法？**

谢锐何：国内的设计公司都在快速成长的过程中，设计水平与国外设计公司的差距正逐渐缩小。而朝着多样化、综合型的方向发展，这是必然的。因为设计的市场份额有限，谁跨界去换位思考，又能在跨界中做出差异，那谁就在市场上能够稳得住神、立得住脚、扎得下根。邦景园林在设计、深化上找到差异化发展的同时，一直强调项目的落地实施性，确保后期深化、实施，使我们区别于其他设计公司，也保证了我们的企业差异性。

**《新楼盘》：您对邦景园林未来有何展望？**

谢锐何：感谢这些年来一直关注、支持邦景园林的单位、朋友，也感谢为邦景园林付出智慧和汗水的每一个团队成员。对于邦景园林的未来，我坚信，只要邦景的团队坚持"真诚、好奇的心"和"为理想而奋斗的精神"，去发现这个时代的问题与机遇，并且愿意为推动"景观的实施和发展"起到的积极作用而努力，定能赋予空间活力与生命，延伸空间独特的价值观；也定能保证邦景园林的独特性，让邦景园林这个企业走得更远。

和丰颖苑

和丰颖苑

和丰颖苑

和丰颖苑

INTERVIEW | 专访

# Differentiation of The Space in Landscape Design

—— Interview with Xie Ruihe, Director and Principal Designer of Guangzhou Bonjing Landscape Design Co., Ltd.

**About the Interviewee**

Xie Ruihe
**Director and Principal Designer of Guangzhou Bonjing Landscape Design Co., Ltd.**

Mr. Xie Ruihe has ever been involved in some famous landscape projects and landscape institutes as the chief designer or project leader. Now he is the director and principal designer of Bonjing Landscape, being in charge of the whole process of projects of different sizes and categories. Skillful in both landscape design and team management, he is qualified in the landscape concept design, detail design, construction drawing design, site engineering and project management.

Till now, Mr. Xie has completed the landscape design and management for hundreds of projects, which include the developments of the famous Vanke Group, Kingold, HOPSON, Poly Group, Times Property, Citic Real Estate, Jiangzhong Group, Hocin Group, etc.

**About BONJING**

GUANGZHOU BONJING LANDSCAPE DESIGN CO., LTD. is a professional landscape company which specializes in landscape planning and design, engineering construction, afforested conservation and seedlings producing. It is a special firm with experienced design team at home and abroad, with professional construction team, with comprehensive resources and sincere service. It focuses on construction technology, project management methods, and finally have established a project management system basing on the construction technology and particular management, in order to ensure that the projects will be finished with high efficiency and high quality. From the beginning to the future, BONJING insists on specialization and branding. According to local conditions, with forward-looking creation concept and differentiated ways of landscaping, it has been creating a poetic landscape environment for everyone, and have been highly praised by the majority of the customers.

**New House:** As a senior landscape designer and executive of a landscape company, would you please talk about your understanding to landscape design?

**Xie:** I feel proud to be a landscape designer, and at the same time I bear the pressure and responsibility it brings.

Landscape design is neither a profession to give impression with simple graphic designs or hand drawings; nor a profession to create space with massive plants and luxurious materials...

The basic function of landscape is to make people experience the style of space. Thus the aim of landscape planning and design is to create spaces of different scales and values. It also tries to well organize the hard and soft landscapes to shape the spaces. What I want to emphasize is that, landscape design is an interdisciplinary subject as well as a long and complicated process of optimization. Many new designers though have the enthusiasm to design, they will be apt to lose patience and give up their initial ideals during the long process of modification. Therefore, there are worries for the future of this industry.

**New House:** Together with your team, you'd ever designed varied landscape projects. In your opinion, what's the role of landscape in residential areas and public areas?

**Xie:** Since its establishment, BONJING has provided solutions and project management for landscape spaces of different scales and categories.

Today's eco environment is changed a lot under the rapid urbanization. With the improvement of the living standards, people begin to long for excellent landscape environment. The real estate developers begin to highlight "landscape" and "ecology" of their products, and municipal project and public space design turns to pay more attention to landscape design. The Report to the 18th National Congress of the CPC proposed the concept of "Beautiful China", emphasizing green landscape and rich life. In my opinion, landscape plays different roles in residential areas and public areas, thus the design should be different accordingly. For the residential area, landscape design should focus on ecology, highlighting the functions and art atmosphere of the spaces. While landscape design for the public area needs to extend the public space and the ecological pattern. Thus in our design, we will provide specific solutions according to local conditions, with forward-looking creation concept and differentiated ways of landscaping. Of cause, it will take more of our time and energy. But we insist to be responsible for the sites, the clients and the works we create.

*New House:* You've advocated the differentiated ways of landscaping, and at the same time put them into some of your practices. So, would you please introduce the the features of these ways to us?

*Xie:* The differentiated ways of landscaping is the design concept for Bonjing.

Differentiation means the pursuit for personality, the respect to different clients, and the insistence to the unique value of each space. Put simply, it's a kind of pursuit to the space effect, to oneself and to the cultural connotation. Without this pursuit, landscaping may become a pure functional design, tedious and stultifying. In BONJING's design, we insist on interdisciplinary collaboration in different stages too create unique and charming landscape spaces of all categories.

*New House:* How does BONJING operate a project? And what's BONJING's view on qualified personnel?

*Xie:* The success of each project is the result of our teamwork. No matter what the scale or location is, we will do field research to know more about the local conditions and communicate with the clients to understand the user experience. The final solution will be made accordingly to create the perfect space.

All the works of BONJING are designed by teamwork, closely connected with the personality, ability and taste of each member. We study together, improve together and create together; never cast away and never give up. So, we've built a great team and we also welcome more talented people to join in us!

*New House:* Would you please share some recent landscape works of BONJING with us?

*Xie:* Recently, BONJING has finished many works and I want to introduce two of them: He Feng Ying Yuan of the bankside of Foshan Nanhai Fairy Lake Resort; phase IV of Tianan Cyber Park Foshan. Both of them are characteristic with our ideas of "differentiation inf space design".

He Feng Ying Yuan: it makes reference to the traditional Chinese gardens and translates them with modern Western landscape elements, presenting innovative landscape spaces. With colorful flowers and plants, traditional Chinese garden culture is mixed with romantic details in Spanish style. Diversified public courtyards are created among flowers and trees to provide beautiful landscapes and leisurely spaces for the owners.

Tianan Cyber Park (Phase IV): the challenge is how to create a differentiated space in this sci-tech park which's already had three phases finished. It's harder to realize creativity in a built modern space than to make innovations in a traditional space. After several times of research and comparison, we decided the this simple but powerful concept. Low-cost and easy-maintenance plant slope serves as the connection of the space. Together with the artistic landscape nodes, it creates an open and free space for the people in high-rise offices.

*New House:* Nowadays, many design firms turn to diversified and compound development. Well, what's your idea about this phenomenon?

*Xie:* Many Chinese design companies grow fast in these years, and the gap of design levels is gradually narrowed between the domestic and foreign companies. Diversified and compound development is necessary. Today, the market share for design is limited, thus who can think over it and make differences accordingly can win the market first. BONJING has found the differentiated development from design and details, emphasized feasibility of the design and made us different from other design companies.

*New House:* At last, would you please talk about the future prospects of BONJING?

*Xie:* We are particularly grateful to our clients and friends who have cared and supported BONJING in these years, and we are also grateful to every team member who spare the wisdom and sweat for BONJING. I believe that, with the philosophy of "sincerity and curiosity" and the spirit of "to struggle for the ideal", BONJING will discover problems and chances to create dynamic and living spaces with unique values. This kind of uniqueness will lead BONJING to go further.

天安数码城

天安数码城

天安数码城

# MODERN SIMPLE LANDSCAPE SPACE WITH FASHION AND TECHNOLOGY ELEMENTS

| Foshan Nanhai Tianan Cyber Park, Phase IV

展现时尚与科技元素的现代简约景观空间
—— 佛山南海天安数码城四期

| 项目地点：中国广东省佛山市 | Location: Foshan, Guangdong, China |
| 开 发 商：佛山市天安数码城有限公司 | Developer: Foshan Tianan Cyber Park |
| 景观设计：广州邦景园林绿化设计有限公司 | Landscape Design: Guangzhou Bonjing Landscape Design Co. Ltd. |
| 占地面积：28 000m² | Land Area: 28,000 m² |

本项目作为天安数码城重点区域，如何创造一种时尚、简约明快又适宜办公人群停留休息的空间，成为贯穿整个设计的宗旨及基调。

项目采用现代造景手法，合理设置各功能区域，通过简练的构图尽可能加大绿化面积，增加空间层次感，加强视野所及的景观感染力。适当布置的休闲平台坐凳使人乐于停留其中，享受属于自己的空间。

景观设计与建筑、室内充分结合，利用干练的铺装线条创造富有节奏感和韵律感的入口空间，为使用者带来新颖而富有变化的景观。

项目整体气势恢弘、简约明快，无论是材料的选择，功能的划分，还是景观小品的设置，在带来强烈视觉冲击的同时，更在布局上保持感官的连续性，并结合天安数码城的文化特征，用当代抽象的设计语言，将空间鲜活地展现出来，赋予空间新的内涵和活力，在满足功能的前提下，更能给人以美的感受。

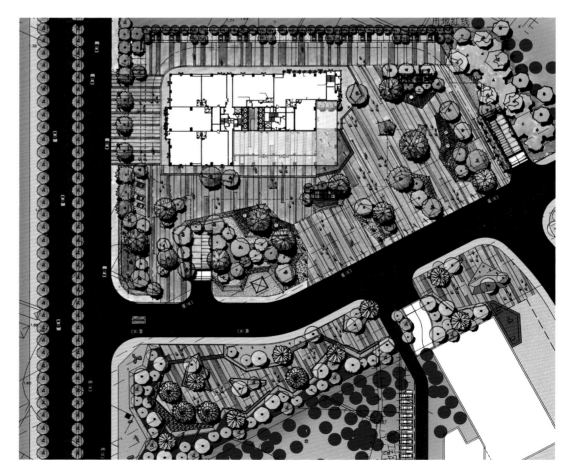

总平面图  Site Plan

NEW LANDSCAPE | 新景观

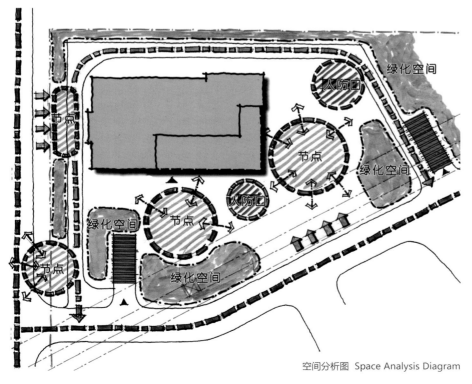

空间分析图  Space Analysis Diagram

This project acts as the key area of Tianan Cyber Park. How to create a stylish, simple and suitable rest space for office staff to stay becomes the designers' aim and keytone throughout the whole design.

The project uses modern landscape design method to set each functional area in a reasonable way that through concise arrangement, it tries to increase more green area, increase the sense of space level and strengthen the attractiveness of the seen landscape. Proper placement of leisure benches makes people happy to stay and enjoy their own private space.

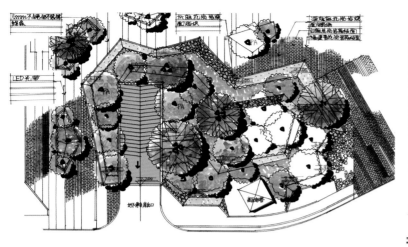
节点平面图一  Node Plan 1

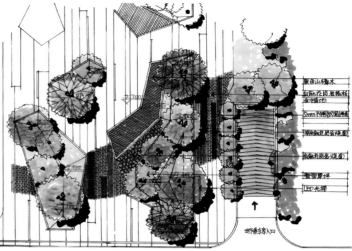
节点平面图二  Node Plan 2

# NEW LANDSCAPE | 新景观

| 公路 | 人行道 | 行道树 | 特色水景 | 特色土坡 | 铺装 |

The landscape design is fully integrated with the building and interior. It adopts concise decorative lines to creating entrance space rich with rhythm sense, providing the users with innovative and changing landscape.

The project looks elegant, concise and lively in whole. The material selection, reasonable layout and landscape settings, not only bring strong visual impact, but also keep continuous feelings in the layout; besides, it combines with the cultural characteristics of Tianan Cyber Park and uses modern abstract design language to show the space vividly, giving the space more new connotation and vitality, which brings people the feeling of beauty under the premise of meeting the functions required.

# CLASSICAL GARDEN WITH MODERN ELEMENTS

| Taigu City Garden Residence, Shenzhen

## 古典园林的现代演绎
—— 深圳太古城花园

项目地点：中国广东省深圳市
开 发 商：宝能地产股份有限公司
景观设计：东大景观设计
项目规模：60 000 m²

Location: Shenzhen, China
Developer: Bao Neng Real Estate Co., Ltd. By Shares
Landscape Design: Dongda Landscape Design
Size: 60,000 m²

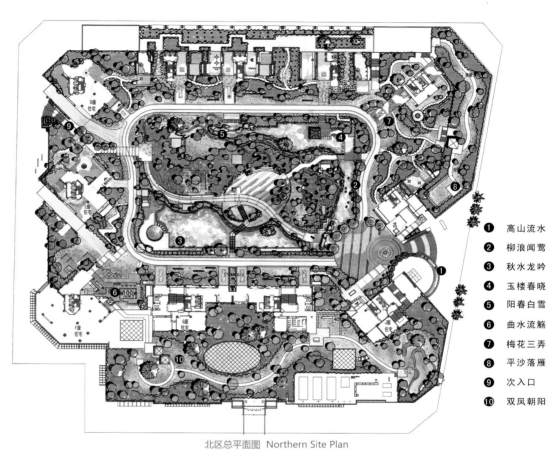

1. 高山流水
2. 柳浪闻莺
3. 秋水龙吟
4. 玉楼春晓
5. 阳春白雪
6. 曲水流觞
7. 梅花三弄
8. 平沙落雁
9. 次入口
10. 双凤朝阳

北区总平面图  Northern Site Plan

# NEW LANDSCAPE | 新景观

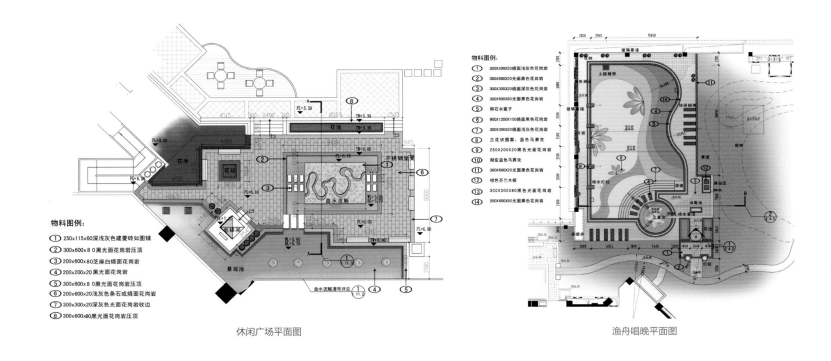

物料图例：
① 250×115×60深浅灰色建菱砖如图铺
② 300×600×80黑光面花岗岩压顶
③ 200×200×80芝麻白烧面花岗岩
④ 200×200×20黑光面花岗岩
⑤ 300×600×80黑光面花岗岩压顶
⑥ 200×600×20浅灰色条石或烧面花岗岩
⑦ 300×300×20深灰色光面花岗岩收边
⑧ 300×600×80黑光面花岗岩压顶

休闲广场平面图

物料图例：
① 300×300×20烧面浅灰色花岗岩
② 300×600×20光面黑色花岗岩
③ 300×300×20烧面深灰色花岗岩
④ 300×600×80光面黑色花岗岩
⑤ 卵石水窝子
⑥ 900×1200×100烧面黑色花岗岩
⑦ 300×300×20烧面浅灰色花岗岩
⑧ 兰花状图案，蓝色马赛克
⑨ 250×200×20黑色光面花岗岩
⑩ 渐变蓝色马赛克
⑪ 300×600×20光面黑色花岗岩
⑫ 棕色芬兰木板
⑬ 300×300×80光面黑色花岗岩
⑭ 200×600×80光面黑色花岗岩

渔舟唱晚平面图

设计理念：设计汲取了中式古典园林的造园精髓，通过现代的造型手法和材料重新演绎，在继承传统园林内在精、气、神的同时，突出现代潮流感及时尚感。设计师于此努力营造高档、现代、新中式音乐自然山水园林式住宅，以现代设计手法及材料表达传统中式园林的形式与意境。

新材料的运用——太古城通过对玻璃、方钢、不锈钢、花岗岩等现代建筑装饰材料在景观构筑物、铺装上的运用，利用传统的中国园林图案和形式的表达，诠释出全新的中式韵味。小区入口处的玻璃跌水假山，以通透的绿玻层层叠加，配合隐藏的灯光设备，表达出传统的中式假山景观。北区曲折的长廊，以整块的弧形玻璃定制顶替换了通常的砖瓦顶，并以现代的钢结构取代了传统的木梁结构。泳池雨廊及池边的屏风装饰将惯用的木质雕刻改为轻钢材料。架空层的装饰景墙也通过钢丝的相互编制沿袭出传统竹编制的装饰效果。铺装材料上则以黑白灰渐变色的花岗岩代替传统的灰砖、青石。园建小品的不锈钢围边，则用简洁的形态重新勾勒出古典家具的复杂线脚。

新颖景观元素造型——对中式传统图案的再次设计加工，成为太古城的另一亮点。运用在架空层门框上的中式窗花图案，在大尺度的冰裂窗格中利用钢丝进行不同方向填充，形成虚实对比，丰富了原有图案变化。地面铺装的碎拼图案，则通过演绎变形及对花岗岩毛面拉丝的材料处理，形成不同的图案纹理，演绎出梅花三弄、玉楼春晓、阳春白雪等中式意境。

新主题的运用——音乐是太古城花园的中心主题，因此设计在小区景点的命名上，也分别以中国传统乐曲命名。北区高山流水、柳浪闻莺、秋水龙吟、玉楼春晓、阳春白雪、曲水流觞、梅花三弄、平沙落雁和双凤朝阳的九大景点及南区碧涧流泉、渔舟唱晚、平湖秋月、阳关三叠、寒鸦戏水、幽兰逢春、三潭印月、渔樵问答的八处景致，各述其境、缓急不一，并通过"水"这一统一元素将各景点——串联，形成包含序曲、开幕、过渡、高潮、小高潮、尾音的一幕大型中式园曲。中国乐典韵律之美与中式传统园林之美和谐交融于太古城的景观艺术设计之中。

# NEW LANDSCAPE | 新景观

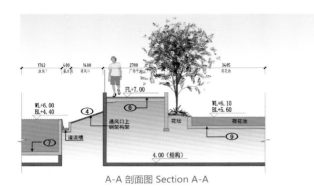

A-A 剖面图 Section A-A

物料图例：
1. 特制弧形玻璃顶
2. 方通外喷灰色氟碳漆
3. 绿色钢化玻璃
4. 钢板刷黑色氟碳漆
5. 300×300×20灰色烧面花岗岩
6. 烧面花岗岩上黑色拉丝
7. 200×200×20光面黑色花岗岩
8. 350×350×80光面黑色花岗岩
9. 200×200×20光面黑色花岗岩

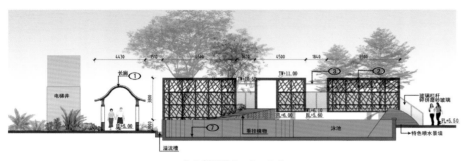

B-B 剖面图 Section B-B

Concept: It is a residential landscape project drawing the quintessence of traditional Chinese garden design with modern modeling methods and materials to highlight the contemporary fashion, which is well integrated into traditional elements and garden architectural forms.

Modern materials – Many modern decorative architectural materials are widely used in this project, such as glass, square steel, stainless steel and granite, to show brand new and special Chinese garden style with traditional garden building forms. A green-glass piled up water-falling rockery is set on the entrance plaza, which gives full express to traditional Chinese rockeryscape by lighting. In the north of this neighborhood unit, a meandering long corridor is covered by a huge piece of arc glass, and the corridor itself is a steelwork. Also, the water-falling gallery and screen decoration on the pool side is built by light steel materials. The decorative wallscape at the empty space is adorned by weaving steel wire. The

pavement of this project is made of black-white granite. All pieces of decorative furniture are embellished by stainless steel edges to reflect the minimalism. Modern materials are one of the main elements in this project instead of traditional ones.

Novel landscape model – The modern design of traditional Chinese garden patterns is another specific in this project. Unique Chinese paper-cut for window decorations on the large-scale ice-crack paned doorframes at the empty space and the allover-patterned pavement both give full express to the integration of modern and tradition.

New landscape theme – Music is the core theme in this project. Each scenery spot here is called by the names of traditional Chinese music, such as The Elegant Music, Orioles Singing in the Willows, The Roaring of Waves in Fall in the north of the site; Blue Waves and Thermal Spring Water, Fishermen Singing the Night Song, Autumn Moon on a Placid Lake in the south part and so on. All these scenery spots are connected one by one as a whole through the unitary element – "water" to form a complete Garden Song. Thus, the project is also a perfect minglement of tradition Chinese music and garden art.

景墙立面图一 Landscape Wall Elevation 1

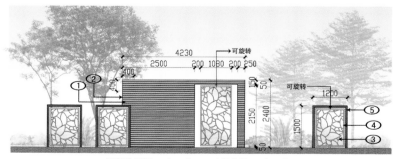

景墙立面图二 Landscape Wall Elevation 2

# MEDITERRANEAN STYLE AND GOLF ENJOYMENT | Lushan Yunling

掬一捧地中海风情 钓一串高尔夫时光 —— 庐山云岭

项目地点：中国江西省共青城市
开 发 商：嘉浩（庐山）房地产开发有限公司
景观设计：深圳市筑奥景观建筑设计有限公司
设计面积：8 899 668 m²

**Location:** Gongqingcheng, Jiangxi, China
Developer: Jia Hao (Mount Lu) Real Estate Development Co., Ltd.
Landscape Design: Shenzhen JOCO Landscape Architectural Design Co., Ltd.
Area: 8,899,668 m²

项目北倚峰峦叠嶂的庐山，东连诗情画意的鄱阳湖，位于素有"鄱阳湖畔的明珠、京九线上的名城"之称的共青城。共青城的南湖湿地生态圈已被列入世界重要原生态湿地名录，山青水秀，风景怡人。此外，项目周边还有五星级的格兰云天大酒店形成功能的互补，交通便利，公共设施相对完善。区域内现状高差为别墅区提供了良好的视线延伸。

项目定位为华中地区最大规模的高尔夫度假别墅群，融合高尔夫球场不事雕琢的粗犷与野趣的设计主题，别墅景观具有深厚的地域风情。设计师根据基地的现状进行规划设计，提出"以人为本、以生态尊"的主题，利用了原本的自然优势"森林与湖"，打造国家级的休闲度假区。设计分为五段景观区域：入口景观节点、滨水观光带、转换景观节点、滨水休闲带、别墅区。

景观设计：设计师从不同地域风格中提取浪漫风情的地中海元素——海岸、阳光、鲜花、陶土等，以绚烂的植物、淳朴天然的材料、地域风情的色彩与风格独具的构筑物来勾勒出一种亲切、柔和、原生的田园风情，演绎回归自然、原生质朴的人生情趣。

园林采用意大利田园风格，以托斯卡纳式农庄为原型，塑造自然悠闲的田园风情。设计上使建筑、园景与自然更好的互相呼应，用比较细致的意大利式的造园手法，体现每个空间的情感，尽显大气华贵。设计师依照起伏的地形，舒展的湖岸线，将自然生态流入各个功能环境。各个空间配合不同的风情景观节点，使艺术性与功能性完美融合。各式古典的花纹与郁郁葱葱的植物，体现自然的幽深，铁艺与风情性的雕塑小品，烘托出优雅气氛。每个景观节点都用合适的自然的石材来体现肌理与古朴典雅。

绿化设计：本土热带植物的大量运用，丰富的植物层次与建筑相得益彰。在各景观区域与节点，采用不同的绿化设计来抒发情感。沿湖景观动感十足，弧线优美的湖泊在阳光下波光鳞鳞，似梦似幻。岸边原有植物种类丰富，且保护较好，设计时予以充分保护和利用。推杆练习场，人们可以在此随性挥洒热力，为了给行走于此周边的人一个视觉上的安全感，在此周边以密植的方式，营造一种高低有序、错落有致、丰富的、自然的、惬意的景观。

总平面图 Site Plan

# NEW LANDSCAPE | 新景观

The project is adjacent to Mount Lu in the north and Poyang Lake in the east, which is located in the Gongqing Town that known as the "Pearl of Poyang Lake and famous city on the Beijing-Kowloon railway line ". Nanhu Wetland Ecological Circle in Gongqing Town is listed into the World Important Ecological Wetland Directory. The five-star Grand Skylight Hotel is at the surrounding area, forming the functional complementation with convenient transportation and perfect public facilities. The local terrace difference provides good sightline extension for the villa area.

The project is defined as the largest-scale golf holiday villa group, fused with rough and wild design theme of the golf course to form villa landscape with strong regional style. Designers go on the design according to the current situation and put forward the idea of "people-oriented and ecological respect". They use the original natural advantage—"forest and lake" to build the leisure and holiday area at the country level. The design divides it into five sections of landscape areas including Entrance Landscape Nodes, Water Front Tourism Belt, Landscape Switch Node, Water Front Leisure Belt and Villa Area.

Landscape Design: The designers extract romantic Mediterranean elements from different regional styles —— coast, sunshine, flowers and clay, etc. Through gorgeous plants, natural materials, local flavor and unique architectural style, it outlines a kind, gentle and native pastoral style, expressing the ideal life, natural and simple.

The landscape design adopts the Italy pastoral style, taking Tuscany farm as a prototype to create the natural leisurely pastoral style. In order to make the architecture, landscape and nature echo with each other, the designers use meticulous Italy landscape design technique to show the luxurious atmosphere of each space. Designers use up-and-down terrain and stretched lake line to integrate natural ecology into all functional environments. Each space, together with different landscape nodes, makes perfect fusion of the artistic and functional qualities. Various classical decorative patterns and luxuriant plant manifest the deep connotation of nature, iron art and sculpture ornaments add more elegant atmosphere. Each landscape node is decorated by the proper natural stone material to reflect its texture and refined flavor.

NEW LANDSCAPE | 新景观

Green Design: The widely use of local tropical plants brings out the best between the plants and surrounding architecture. In different region and nodes, it uses the different greening design to express different feelings. Along the lake, there is dynamic landscape. The lake, with beautiful arc line, sparkles under the sun and feels dreamy and dramatic. The rich original plant species get good protection and be full protected and utilized in the design. In the golf swing courses, people can do exercises whenever they like to. In order to make a visual sense of security for people who walk along the golf swing courses, it makes condensed planting around to create landscape that well-proportioned, rich, natural and comfortable.

FEATURE | 专题

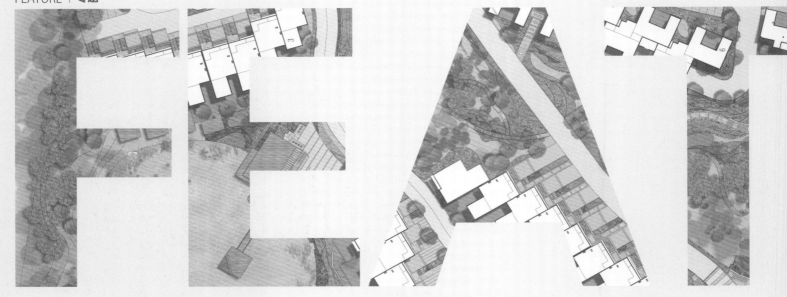

# 中西部崛起

## 专题导语

近年来，在国家宏观调控政策下，中西部地区的发展速度明显加快。随着中西部地区崛起战略的推行、东部沿海地区产业的转移以及房产调控的日趋严厉，越来越多的开发商转向了土地和市场均被看好的中西部二、三线城市。中西部地区房地产的发展起步稍晚，起点较低，但后发优势明显。在建筑设计上，中西部地区更加突显地域特色，因地制宜，打造让人眼前一亮的产品；同时注重与环境的紧密融合，使自然与住宅浑然一体，打造生态、宜居型住宅楼盘；此外中西部楼盘在空间结构、户型采光以及景观塑造等方面也体现着南北差异。由于中西部地产开发整体处于初步阶段，其发展过程中依然存在诸多的问题，如何充分利用地区优势，打造更多符合需求的住宅，仍需深入探索。本期专题，我们将与您一起分享几个具有代表性的中西部楼盘，共同探寻中西部楼盘设计方面的新思路。

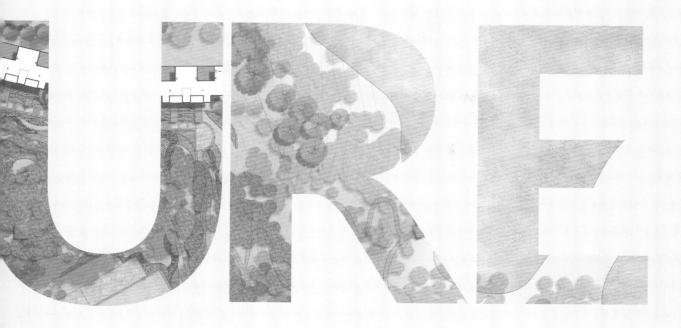

# Introduction

Recent years, central and western China developed rapidly under the state's macro-control policy. With the promoting of the central and western development strategy as well as the increasingly strict real estate regulations, more and more developers turn to the second and third-tier cities in the central and western China to find a further development. Though developed later, the real estate industry in these regions enjoys great advantages: the architectural design here emphases more of the local characteristics to present eye-catching products; meantime, it pays attention to combine with the surrounding environment, trying to creating ecological and comfortable housing projects; in addition, the space, interior layout, lighting and landscape of the houses in the central and western regions are also designed to be different from that in the other regions. Since it is at the preliminary stage of the development strategy, there are still some problems to be solved, for example, how to take advantages of the existing resources, how to design more housing projects that satisfy local demands, etc. In this issue, we will introduce some housing projects in the central and western China to you, and try to find some new ideas for housing developments in these regions together.

# 重庆隆鑫地产（集团）有限公司
# LONCIN LAND (GROUP) CO., LTD.

**企业简介**

重庆隆鑫地产（集团）有限公司是一家总部设在重庆的全国性大型房地产企业，是国务院发展研究中心等权威机构联合评定的中国房地产百强企业，拥有国家一级房地产开发企业资质、国家一级物业管理企业资质。作为中国最具成长潜力的房地产企业之一，隆鑫长期致力于提升产品及服务品质，工程项目四揽全国詹天佑大奖，物业管理三届蝉联全国物业服务满意度十佳，产品及物业服务品质稳居全国领先地位。

**产品与服务**

历经9年时间的潜心积淀，隆鑫业务领域已涉及地产开发、酒店经营及物业管理三大板块，全国开发的各类项目近30个，开发面积总计逾300万$m^2$，形成了精品高层系列、风情洋房系列、城市别墅系列、旅游地产系列、城市综合体系列五大专业产品线，具备了花园洋房、高层、别墅、酒店、写字楼等多种物业类型的丰富开发经验。根据战略部署，隆鑫在深耕本土的同时，已成功进入成都、太原、海南、云南等区域市场，储备用地逾万亩，在建项目近200万$m^2$，资源能力与发展后劲居于行业前列。

**Profile**

As a big real estate enterprises based in Chongqing City, Loncin Land (Group) Co., Ltd. Ranked Top 100 China Real Estate Enterprises by authorities like the Development Research Center of the State Council (DRC), having the state-level-A real estate development and management qualification. Regarded as one of the most potential real estate enterprises, Loncin is always dedicated in improving the quality of its service and products. Its projects have won Zhan Tianyou Award for four times, its real estate management is fully accepted by clients for three consecutive years, and the quality of its service and products keeps top in China.

**Product and Service**

With nine years of growth, Loncin's business has covered real estate development, hotel management and properties management. With thirty projects covering a total area of more than 3,000,000 $m^2$, Loncin's products can be classified into boutique high-rises, garden houses, urban villas, tourism estates and urban complexes. Now, it has rich experience in the development of garden house, high rise, villa, hotel, office and so on. According to its development strategy, with emphasis on its headquarter – Chongqing City, it's also entered into Chengdu, Taiyuan, Hainan, Yunnan, etc. With 2,000,000 $m^2$ projects under construction, it becomes one of the leaders in real estate industry.

# 中化方兴置业（北京）有限公司
# SINOCHEM FRANSHION PROPERTIES (BEIJING) LIMITED

**企业简介**

中化方兴置业（北京）有限公司主营房地产的开发、建设及商品房的出售、出租和物业管理服务，其投资方方兴地产（中国）有限公司是中国中化集团公司在地产酒店领域的旗舰企业。公司在项目开发实践中总结、探索、创建了一套可移植、可复制、有特色的住宅类项目开发管理体系，并努力打造一支高素质、专业化的高端住宅类产品开发管理人才团队，形成了其高端住宅产品品牌和市场知名度及影响力，为方兴地产高端住宅业务持续发展做出积极的贡献。

**产品与服务**

公司致力于将其发展成为包括高端居住物业、城市高档公寓、体育公园、小学等在内的融人文、休闲、教育和时尚为一体的城市地标级高端居住综合体，并以方兴地产五星级酒店管理服务模式，为客户提供高效、优质的管家式服务，成就名流人士世界级纯高端居住理想。目前，公司正在开发建设的北京市朝阳区广渠路15号地项目被称为"CBD核心区域绝版黄金地块"。同时，公司秉承中化集团"做人：诚信、合作、善于学习；做事：认真、创新、追求卓越"的文化理念，积极开拓新增土地储备工作，拓展地产业务的覆盖范围，并努力为客户打造满意程度高的精品项目，为股东创造丰厚的投资回报。

**Profile**

Sinochem Franshion Properties (Beijing) Limited is dedicated in the development, construction, sales, renting and management of properties. Its investor – Franshion Properties (China) Limited is a member of Sinochem Group, focusing on real estate and hotel development. With years of experience, it established a transplantable, reproducible and characteristic system for residence development and management. And it's also tried to build a talented team of high education and high professionalism for high-end residence development and management. Now it has a great reputation in this market, which will also contribute to its future development.

**Product and Service**

The company devotes to develop landmark complexes which integrate high-end residential residences, luxury apartments, sports parks, primary schools and so on. With its standard management model for five-star hotels, it tries to provide clients with high-efficient and excellent housekeeper-style services. Recently, its new project at No. 15 Guangqu Road of Chaoyang District, Beijing which is regarded as the "golden block in CBD" is under construction. Meantime, with Sinochem's cultural philosophy of "honesty, cooperation, learning, earnest, innovation, and excellence", it tries to increase the land purchases, expand its business, provide clients with boutique projects and create more profits for the shareholders.

# ECOLOGICAL, AFFORDABLE AND ECONOMICAL UPPER-MIDDLE STANDARD RESIDENTIAL COMMUNITY

| Inner Mongolia Qiushi Jing Mayfair Residential Clubhouse

绿色、适用、经济型的中高标准居住社区
——呼和浩特秋实璟峯汇

项目地点：中国内蒙古自治区呼和浩特市
建筑设计：北京维拓时代建筑设计有限公司
总用地面积：153 022 m²

Location: Hohhot, Inner Mongolia, China
Architectural Design: Beijing Victory Star Architecture & Civil Engineering Design Co., Ltd
Total Land Area: 153,022 m²

## 项目概况

该地块位于呼和浩特市赛罕区，西邻正在建设的锡林郭勒南路，北侧为赛罕西街，南至陶利东街，东临金桥路。总用地面积153 022 m²。基地邻近市中心，地理位置优越，有多路公交车相联系，交通便捷；周围紧邻内蒙高教区，环境优美，拥有良好的人文环境，升值空间巨大。

## 规划布局

社区规划以建设"绿色、适用、经济"型的中高标准居住社区为理念，在经济性与高品质的共同追求中寻求最佳契合点。项目规划采用"两轴、两点、多中心"的构架，多角度细节设计提升其品质。在社区规划中注重体现品质节能环保等理念，保温性能达到65%节能标准，应用中水系统，雨水收集和节电设备。在诸多方面反映时代的发展和技术手段的进步，使之成为符合现代标准的宜居社

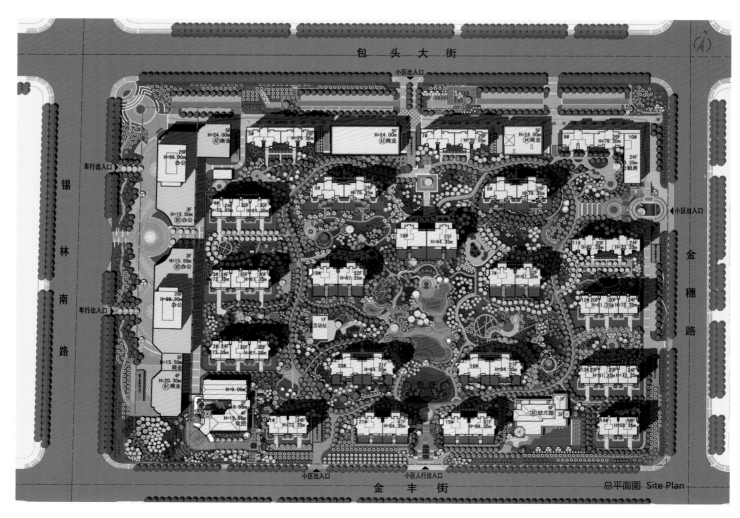

总平面图 Site Plan

区。公建配置设计适合其社区文化服务设施：增加老年活动设施，设立便民的物业服务大厅，沿街适当增加商业设施，以服务社区内外的人群。

**建筑设计**

该项目建筑单体形式较为丰富，包括舒适型高层，豪华型高层及紧凑型的高层三种形式。建筑立面的设计与项目定位相匹配，利用建筑自身形体形成修长、优美的轮廓，并将地中海风格融入其中，建筑色彩温暖柔和。外立面材质考虑到造价因素，以涂料为主，在低层和局部使用石材，以提高小区品质，整体造型与目标客户群的品位需求相一致。建筑体量高低错落，变化多样，避免单调感。顶层采用坡屋面，结合退台处理手法，丰富建筑的第五立面。强化单元入口及细部设计，在经济性的条件下创造温馨的社区环境。

**户型设计**

在户型设计上，对户型进行精细化设计，从人与家具的尺度出发，去除以往户型中过剩的面积，达到户内动线最合理。

**景观设计**

利用入口和道路交叉形成的有利空间设置景观节点，并利用竖向坡地塑造立体绿化。同时结合交通规划，将一部分车行环路设计为硬地铺砌的景观式道路，通过物业管理控制，仅供少量车辆在特殊情况下使用，在平时作为景观步道，实现人车分流，保持小区主路顺而通畅的特点，形成小区路——路团路——宅前路的层级结构。

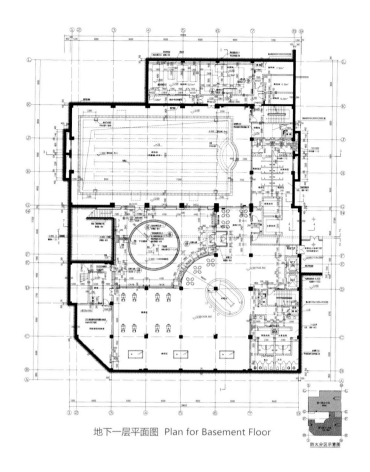

地下一层平面图　Plan for Basement Floor

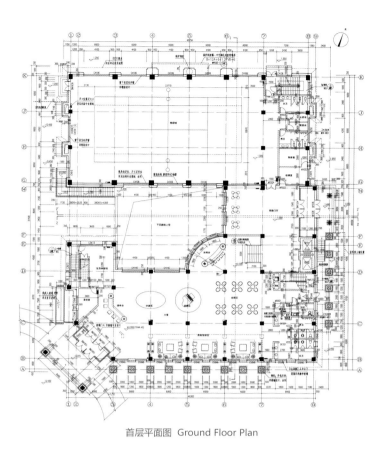

首层平面图　Ground Floor Plan

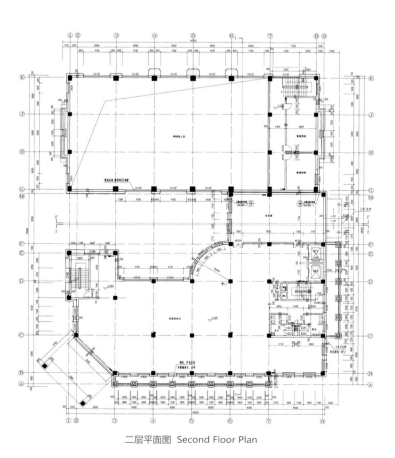

二层平面图　Second Floor Plan

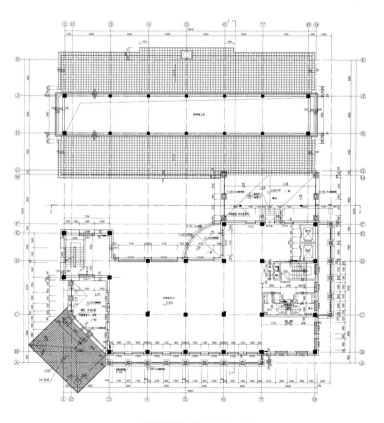

三层平面图　Third Floor Plan

NEW HOUSE _ 071

FEATURE | 专题

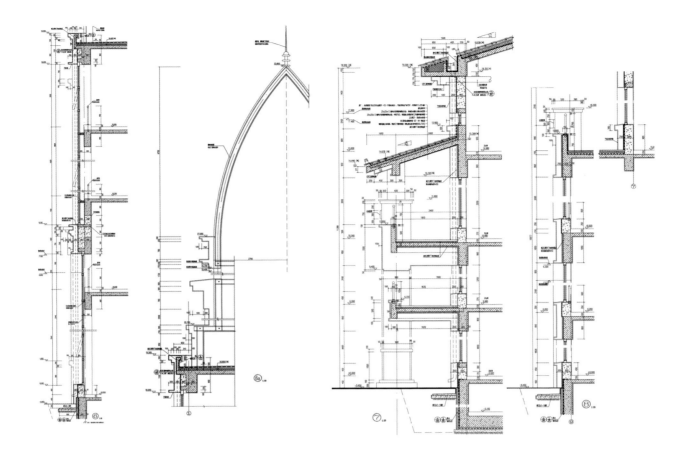

## Overview

This plot is within Saihan District in Hohhot city, west is nearby South Xilingol Road, north is adjacent to West Saihan Street, south is towards to East Taoli Street and east faces to Jinqiao Road. The total land area is 153,022 m$^2$. Due to close the downtown, the foundation is well located and with convenient transport lines, around which is higher education district of Inner Mongolia, possessing exquisite environment as well as sound cultural environment and having great appreciation value.

## Planning

Based on the concept of building " Ecological, affordable, Economical " upper-middle standard residential community, the planning seeks for the best conjunction point in the pursuit of economics and high quality, and the detail design improves the quality through the appliance of "two axis, two points, multiple centers" structure. What is more, the planning design is focus on concepts like quality reflected, energy conservation and environment protection etc, the thermal insulation properties reach energy-saving standard to 65% together with other energy-

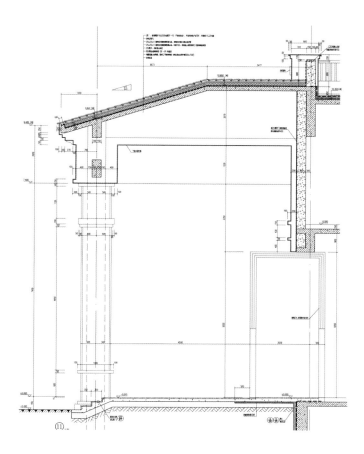

# FEATURE | 专题

efficient equipment. This Planning has reflected times development and technology improvement so as to build up a livable community fit for contemporary standards . The public facilities construction is designed to be suited to the community service, for instance, adding supports facilities for the elder, building up property service hall for residents and increasing members supplying service for commercial zone.

**Architecture Design**

Buildings form is quite variety, including comfortable type high-rise, luxury type high-rise and compact type high-rise. The facade design matches the project position, taking advantage of building itself to create a slender but elegant outline and infusing with Mediterranean style so as to make the color tone of buildings soft.

Due to the cost, the material of exterior facade is painting-based, while the low-rise and parts uses stones in order to improve the community quality. The overall shaping design is corresponding to target customers requirement. Building volumes is various to avoid tedium. The top floor is pitched roof combined with terraced approach, which enrich building's fifth facade. The enhancement design of the entrance and details makes a harmonious community environment under economical condition.

**Housing Design**

The detailed design for units start from the angle of human and furniture, wiping out the excess space to get the indoor generatrix suitable.

**Landscape Design**

The utilize of the entrance and the road cross into a landscape node in favor for spatial arrangement. Also making the use of the vertical slope to construct vertical planting. Combined with transportation planning, part of driving circuit is designed to be landscape loop with hard court. By means of Property management control , this road could only be used by few cars under exceptional circumstances. Usually this road is used as landscape trail in purpose of splitting flow of people and vehicles and keeping the main road in community clear, producing a unique hierarchy.

# URBAN GARDEN LIFE COMBINING CHINESE AND WESTERN ELEMENTS | Chengdu Nanjun Qiyingli

中西结合畅享都市内的庄苑生活
—— 成都南郡七英里

| 项目地点：中国四川省成都市 | **Location:** Chengdu, Sichuan, China |
| 开 发 商：成都裕鑫房地产开发建设公司 | **Developer:** Chengdu Yu Xin Property Development and Construction Co., Ltd. |
| 建筑设计：北京翰时国际建筑设计咨询有限公司 | **Architectural Design:** Beijing A&S International Architectural Design and Consulting Co., Ltd. |
| 建筑面积：85 355m² | **Floor Area:** 85,355 m² |

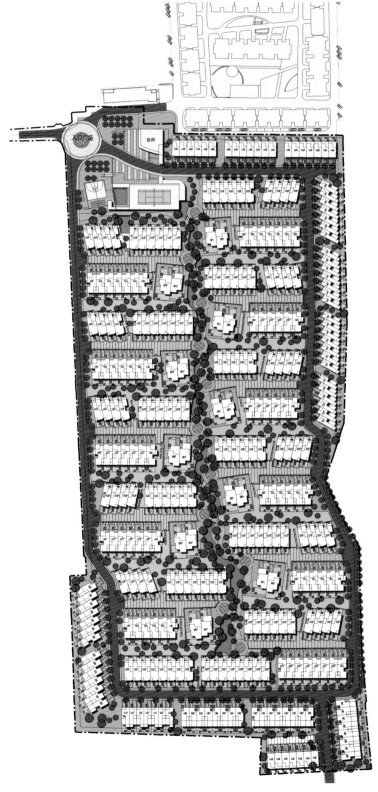

总平面图 Site Plan

FEATURE | 专题

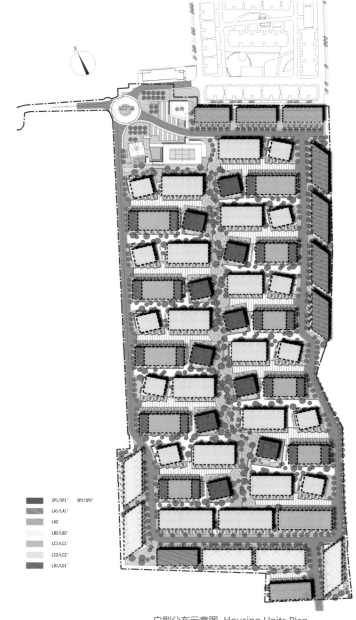

户型分布示意图 Housing Units Plan

**项目概况**

该项目位于成都市以南人民南路沿线，西邻成仁公路。项目用地面积为99 693.8 m²，规划总建筑面积85 354.8 m²。

**规划布局**

规划整体布局借鉴欧洲小镇住宅的形式，同时融入中国传统庭院理念，通过建筑体量围合，形成相对私密、尺度宜人的组团院落，营造既保持传统文化又独具西方舒适性的生活氛围。组团绿化向中心水系渗透，将每个组团包围在环状绿岛中央。

**户型设计**

项目户型设计极大的满足了使用的合理性及舒适性，每户均享有前后庭院及围合式内庭院，在户内形成露天空间，各种不同庭院空间的设计，继承川西文脉，同时又加以创新，以邻里空间的塑造为主题，创造都市内的庄苑生活。

立面图 1  Elevation 1

立面图 2  Elevation 2

立面图 3  Elevation 3

立面图 4  Elevation 4

FEATURE | 专题

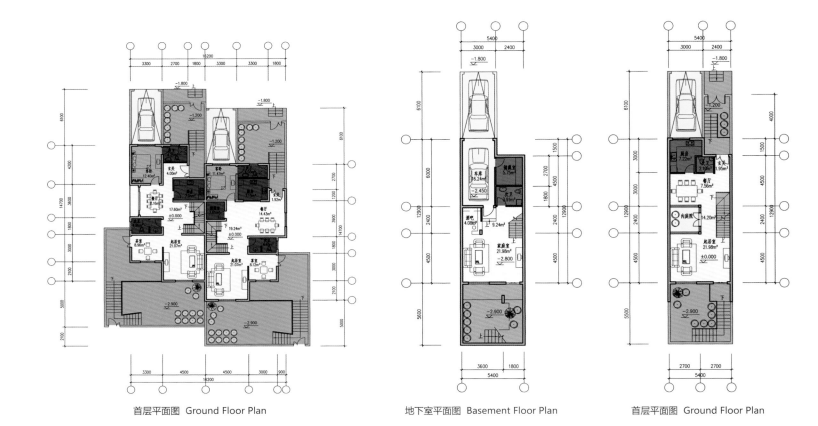

首层平面图 Ground Floor Plan　　　地下室平面图 Basement Floor Plan　　　首层平面图 Ground Floor Plan

FEATURE | 专题

## Overview

This project is located along Renmin South Street and west to Chengren Road. It has a land area of 99,693.8 m² and a total planned area of 85,354.8 m².

## Planning

The planning of this community is according to the type of European town and adds in Chinese traditional courtyard. It combines Chinese traditional culture and western comfortable life style through the building collocation to form private and pleasant courtyard group. The greening system pervades to the central water system, which encircles each group in the center of the orbicular green islands.

## Housing Design

The housing design is reasonable and comfortable, and each room has a front garden, a back garden and an inside one, which creates large indoor courtyard space. The design of the courtyard space carries on the Chuanxi culture and also brings forth new ideas with the theme of building the neighborhood space, creating urban garden life.

# GARDEN RESIDENTIAL COMMUNITY IN NEW BAYU (CHONGQING) STYLE

| Chongqing LONCIN Hong Fu

新巴渝风的花园式居住社区
—— 重庆隆鑫鸿府

| | |
|---|---|
| 项目地点：中国重庆市 | **Location:** Chongqing, China |
| 开 发 商：重庆隆鑫地产(集团)有限公司 | **Developers:** Chongqing LONCIN Real Estate (Group) Co., Ltd. |
| 建筑设计：ZNA泽碧克建筑设计事务所 | **Architectural Design:** Zeybekoglu Nayman Associates |
| 设 计 师：董晖 | **Designer:** Dong Hui |
| 甲方项目经理：冯凌 | **Project Leader:** Feng Ling |
| 场地面积：151 308.5 m² | **Land Area:** 151,308.5 m² |
| 建筑面积：367 617 m² | **Floor Area:** 367,617 m² |

**项目概况**

该项目用地位于重庆市渝北区农业园区内，紧邻机场道路（201国道），西面是渝北区经开大道，北面距离机场约6 km，南面距江北观音桥约8 km。项目总占地面积373 600 m²。项目定位为花园式生态居住社区。

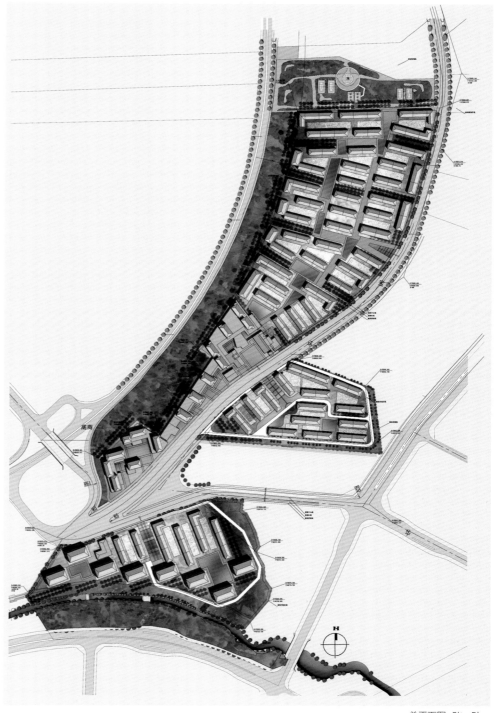

总平面图 Site Plan

C户型南立面图 South Elevation of Type C Unit

C户型东西立面图 East And West Elevations of Type C Unit

**规划布局**

项目划分为A1、A2两部分。A1区主要以水平方向排布的多层住宅为主，规划避免将建筑的主采光面朝向机场路，注重加长快速路一侧的城市观感，用建筑语言同时结合山地地形，刻画出动感活跃、层次丰富的天际线。园区内建筑组团讲求庭院空间的设计，创造花园式宜居社区。

A2区沿街布置展开面长度约500 m的商业配套设施，既服务社区，也辐射至两江新区周围10 000 000 m² 的经济发展带。商业区规划两个面向公众开放的商业围合城市广场，并由一条街道型商业街串联，商业内容丰富，建筑形式多样，有效地提升了此区域的生活品质。此外，合理布置各种配套服务设施，使其功效最大化，充分考虑消防、停车、幼托、物管、社区商业用房、变电房等设施的布置。

**建筑设计**

建筑造型采用灰色调的中式风格建筑，通过简洁的建筑线条、庄重的建筑色彩和独特的风格，探讨新巴渝风的朴素建筑，使整个建筑群体呈现出与当地自然、人文共荣的和谐韵味。同时又以较为得体的抽象画的处理和协调的尺度比例与体量组合，体现一定内在的建筑美感。

住宅楼外景 Landscape View Outside Residential Units

FEATURE | 专题

会所外景 Landscape View Outside Club

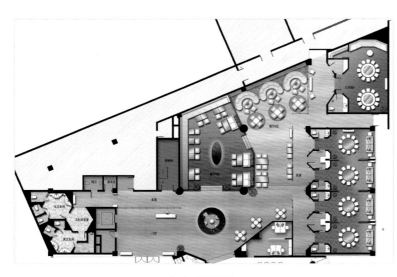

会所一层平面图
Plan of the 1st Floor of Club

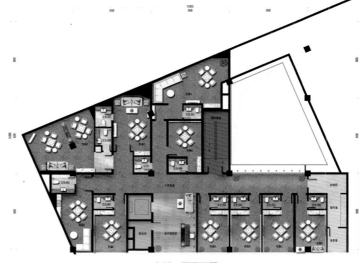

会所二层平面图
Plan of the 2nd Floor of Club

**户型设计**

户型设计方正实用，所有卫生间均能对外开窗，符合当今强调健康住宅的潮流。功能结构和分区设计合理，厨房、入口和餐厅相结合，卫生间和卧室划分一处，避免动线交叉。入口处设置灰空间作为室内外过渡空间。此外，简化结构，整理建筑外型，尽量墙柱拉齐，外墙方整，以利于结构体系规整，节省造价。每户都有前后花园，客厅上空，主卧室带大露台。

**景观设计**

项目采用软硬兼顾的景观设计，即建筑化的景观（硬景）和自然化的景观（软景）合理结合，既考虑经济性又能达到很好的效果，除了功能因素之外，还可供观看、欣赏，今后做专业景观时也无需太多额外的小品出现，建筑本身已经能深深地把建筑和景观连成一体：景观是建筑空间的延伸，建筑演绎景观。因建筑的风格、造型是总的设计思路控制下的产物，从而整个景观的设计主题也与建筑规划主线相一致。

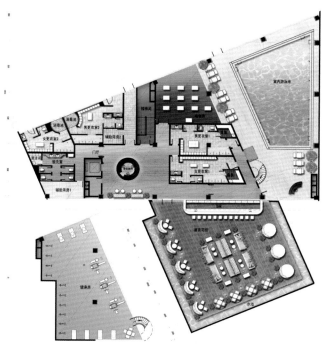

三层平面布置图 3rd Floor Plan

售楼处外景 Landscape View Outside Sales Center

**Overview**

The project is located in the agricultural park of Yubei District in Chongqing. It is adjacent to the Airport Road (201 National Highway), west to the Jingkai Road in Yubei District, about 6 kilometers from the airport in the north and 8 kilometers from Jiangbei Guanyin Bridge in the south. The project covers a total area of 373,600 $m^2$. It has been positioned as a garden-type ecological residential community.

**Planning**

The project is divided into A1 block and A2 block. A1 mainly includes multi-storey residential buildings arranged horizontally. The planning of A1 avoids the main lighting surface towards the Airport Road and pays attention to enhance the city impression of one side of the expressway. Combined with the mountainous terrain, it uses architectural language to build the dynamic, active and rich-layer skyline. The design of the building group pays attention to the courtyard space to create a garden livable community.

A2 block is arranged along the street and the length of commercial facilities is about 500 m, which not only services for the community, but also brings vigor to the 10,000,000 $m^2$ economic development zone of the Liangjiang New Area. The commercial block plans two public commercial plazas connected by a street commercial street with rich commercial content and various building forms, effectively promoting the life quality of that region. In addition, the rational layout of various service facilities maximizes its functional ability that it fully considers the arrangement of the fire protection, parking, kindergarten, property management, commercial housing and transformer facilities.

**Architectural Design**

The project modeling adopts Chinese gray color-tone style. Through simple construction lines, solemn architectural color and unique style, it tries to explore the new simple Bayu architecture to make the whole building groups become harmonious with the local nature and people. In addition, the reasonable scale and combination of volume reflect some kind of intrinsic architectural beauty.

**Housing Design**

The housing design is practical and appears in side that all the washrooms have windows open outwards, which accords with the emphasis of healthy living environment. Its functional structure and independent partitions are arranged in a reasonable way. The kitchen, entrance and dining room combine tightly, while the bathroom and bedroom site on another side, thus avoiding disturbing with each other. Gray space from the entrance is the transitional belt for the indoor and outdoor space. In addition, it tries to simplify its structure, aline its wall column, square its exterior wall to facilitate a regular structure system and save the construction cost. Each household has gardens in front and behind, and the master bedroom keeps with a large balcony.

**Landscape Design**

The project combines the architectural landscape (hard view) and the natural landscape (soft landscape) in a reasonable way, which seems both economical and efficient. Apart from its functional elements, it can also be used for viewing and appreciating, and needs not much more extra pieces to do professional landscape in the future. The building has been united tightly with the landscape: the landscape is the extension of architectural space, and the architecture interpretation of landscape. The architectural style and the building shape root in the entire design conception of project, so the design theme of the landscape is also consistent with the building planning.

住宅楼阳台景观 Balcony View

典型户型首层上跃平面图
Ground Floor Plan of Standard Unit

典型户型首层下跃地下一层平面图
Basement Floor Plan of Standard Unit

会所室内 Club Interior

住宅室内 Residence Interior

# NATURAL RESOURCES MAKING FIRST VILLA COMMUNITY IN CHENGDU | Philippe Hills Villa

原生资源,造就成都第一纯别墅区 —— 成都三盛·翡俪山

| | |
|---|---|
| 项目地点:中国四川省成都市 | Location: Chengdu, Sichuan, China |
| 开 发 商:三盛地产 | Developer: San Sheng Real Estate Group |
| 占地面积:61 000 m² | Land Area: 61,000 m² |
| 建筑面积:72 000 m² | Floor Area: 72,000 m² |
| 容 积 率:0.70 | Plot Ratio: 0.70 |
| 绿 化 率:40% | Greening Rate: 40% |

## 项目概况

该项目所在的牧马山区域，有长达10年左右的别墅开发史，拥有成都第一个高尔夫球场，被誉为"成都首席别墅区"。随着交通配套的改善和产品的强势定位，翡俪山打破了北京温榆河、上海佘山、深圳香蜜湖主导的中国中央别墅区版图格局，成为与之并驾齐驱的西部第一中央别墅区。

## 规划布局

三盛·翡俪山享有牧马山约1 333 333 m² 原生山林、顶级高尔夫球场、在建牧马天堂与牧山湖，还将拟建瞿上城市公园等尖端配套资源。其绝好的自然浅丘坡岭地貌和山水资源，以及容积率小于1的区域规划，使牧马山成为高端低密度别墅的代言品牌。

## 建筑设计

项目建筑对位高端人士的需求，在奢侈主义的大潮中，抛弃奢豪的外在，转而寻找精致的内在，将适度、适合的原则贯穿于别墅区的设计中。独有的全主人生活系统，充分尊重每一位家庭成员的私密，整层豪华主人空间，更将纯别墅风范演绎到极致。

## 户型设计

联排别墅面积约260~310 m²，独栋别墅面积在510 m²左右。而户型的设计尤其关注得房率，丝毫不浪费一寸空间；户户采光，通透明亮，创造出别具一格的，性价比高的纯别墅产品。

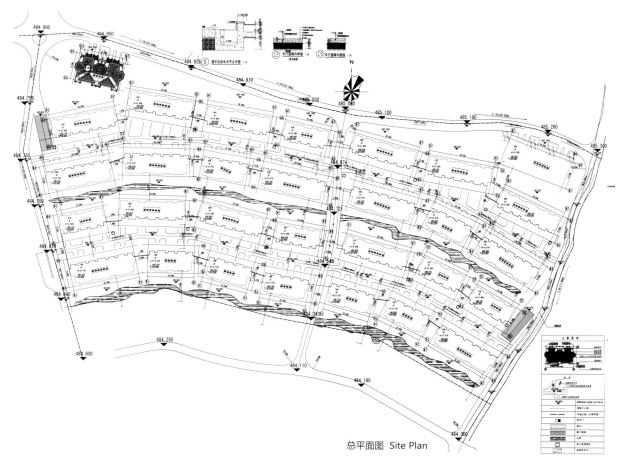

总平面图 Site Plan

立面图 1 Elevation 1

立面图 2 Elevation 2

**景观设计**

项目借助牧马山丰富河系资源，坐拥三河两湖一渠，更动用精妙设计与高昂成本，最大限度地保留山间的原生树和老竹林。在滨河林荫道上有胸径达40 cm以上的老香樟树，形成业主进入别墅区的第一重礼仪。项目建筑精心引水，每一栋别墅都能最大限度的与水亲近。项目别墅环湖而建，数千平方米的中央湖，独辟出弥足珍贵的别墅景观。该项目临河的建筑，特意设计有八角塔，而向山的建筑，设计有大露台。别墅中还设计了全采光的HOME-PARTY花园层，种植的花卉能全方位接受阳光和雨露，这让精美花园层成为举办HOME-PARTY的绝佳场所。

立面图 3  Elevation 3

立面图 4  Elevation 4

立面图 5  Elevation 5

立面图 6  Elevation 6

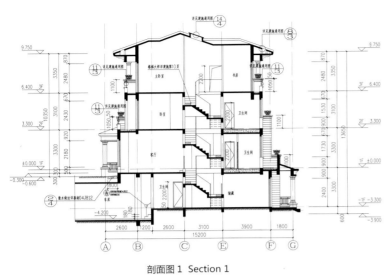

剖面图 1 Section 1

剖面图 2 Section 2

**Overview**

Mumashan Area where the project locates in has been developing villas for up to 10 years, holding the first golf course in Chengdu and being called as "Chengdu Top Villa Residential Area". Along with the improvement of transport facilities and strong product positioning as well as being top central villa residential area in West China, Feili Mountain breaks China central villa's domain pattern led by Beijing Weiyuhe, Shanghai Sheshan, and Shenzhen Xiangmihu and keeps pace with them.

**Planning**

Feili Hills Villa possesses around 1,333,333 m$^2$ forest of Mumashan Mountain, top golf course, Muma Paradise and Mushan Lake under construction, where proposes to build top supporting facilities like Qushang City Park. Due to its outstanding terrain and natural resources as well as regional planning (ratio<1), Feili Hills Villa turns to be the representative of high-end low-density residential villas.

**Architectural Design**

Aimed at requirements of high-end individuals, architectures of this project seeks for the delicate interior instead of luxury and extravagant appearance, and applies appropriate and suitable principle through the villa area design. Furthermore, the unique live system fully respect the privacy of each family member, and the gorgeous master room covering the whole floor completely interprets the charming of villa.

**Housing Design**

In respect to unit type design, the townhouse covers occupies around 260-310 m$^2$ while the single house covers an area around 510 m$^2$. Particularly the unit type design focuses on space availability, trying not to waste any space. Each unit get adequate lighting, bright and transparent, creating a special villa with highly cost effective.

FEATURE | 专题

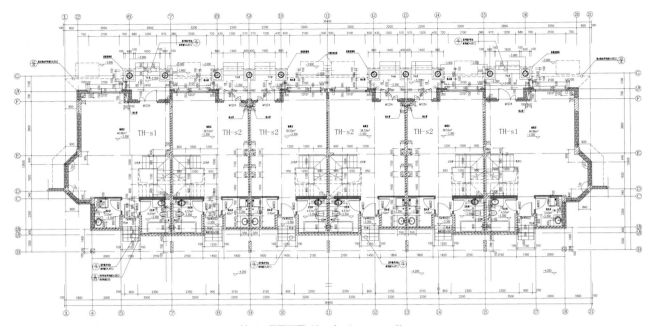

地下一层平面图  Plan for Basement Floor

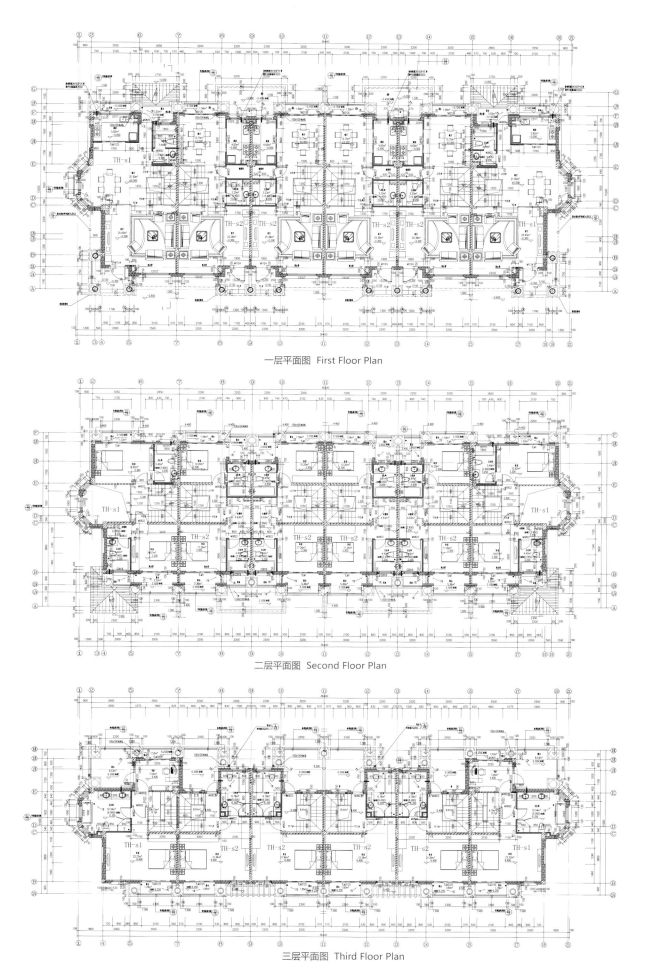

一层平面图 First Floor Plan

二层平面图 Second Floor Plan

三层平面图 Third Floor Plan

# FEATURE | 专题

**Landscape Design**

By virtue of rich river resources in Mumashan Mountain and sitting nearby three rivers, two lakes and one canal, this project takes full advantage of delicate design and high cost to maximally reserve the woods and old bamboo forest. The old camphor tree with width over 40cm in marina boulevard means the beginning of residents stepping into the villa area. This project is designed to transport water meticulously that each villa could get close to water as much as possible. Buildings are built around the lake which covers thousands square meters area , creating a precious villa landscape. The buildings facing the river is particularly design with octagonal pagoda, while those facing mountains is design with wide terrace. What is more, there is a home–party garden  floor with comprehensive lighting ,on which plants could fully under the sun and the rain so as to make it a perfect place to hold a home–party.

NEW CHARACTERISTICS | 新特色

# UPSCALE RESIDENTIAL COMMUNITY WITH SPECTACULAR MOUNTAIN VIEWS AND ROMANTIC SEASCAPE

Heheng Urban Plaza (Haishan Avenue No. 1)

尽享巍峨山景与浪漫海景的高档住宅区——深圳和亨城市广场

# NEW CHARACTERISTICS | 新特色

项目地点：中国广东省深圳市
开 发 商：深圳市创建业房地产开发有限公司
　　　　　深圳市暗径股份合作公司
建筑设计：深圳合大国际工程设计有限公司
用地面积：19 472.34 m²
总建筑面积：122 412 m²
住宅建筑面积：77 894 m²
商业建筑面积：22 062 m²
会所建筑面积：1 300 m²

**Location:** Shenzhen, Guangdong, China
**Developer:** Shenzhen Chuangjianye Real Estate Development Co., Ltd
　　　　　　Shenzhen Anjing Holding
**Architectural Design:** Shenzhen HD Design
Land Area: 19,472.34 m²
Total Floor Area: 122,412 m²
Residential Floor Area: 77,894 m²
Commercial Floor Area: 22,062 m²
Club Floor Area: 1,300 m²

**项目概况**

　　和亨城市广场属盐田区旧村改造项目，是一个享受梧桐山国家森林公园山景、明思克航母海景，周边配套齐全，居住氛围浓厚的高档住宅小区。

**规划布局**

　　该项目位于深圳市盐田区沙头角海山路东侧，北临香径东路，南临海都花园，东邻盐田区福利住宅区海山居，西邻盐田行政文化中心、书城。项目规划6栋13~30层住宅，其中地下室2层，共有近700套单位，660个停车位。

**户型设计**

　　户型主要包括40 m²左右的1房、50~80 m²的2房、120 m²左右的3房和140~160 m²的4房。

NEW CHARACTERISTICS | 新特色

## Overview

As a reformation project of an old village in Yantian District, Heheng Urban Plaza is a high-end residential community surrounded by the mountain scenes of Wutongshan National Forest Park and the seascape of Minsk. It is equipped with completed supporting facilities to be a community of strong dwelling atmosphere.

## Planning

The project is located in the east of Haishan Road, Shatou Jiao, Yantian District on Shenzhen, with Xiangjing East Road to its north, Haidu Garden to its south, Fuli Shanhai Dwellings to its east and Yantian administration and culture center and the bookstore to its west. It occupies 6 buildings of 13 to 60 floors, including two basement floors, 700 unit houses and 660 parking spaces.

## Housing Design

The house plans of the projects include single-room houses of about 40 m$^2$, double-room houses of 50 to 80 m$^2$, three-room houses of 120 m$^2$ and four-room houses of 140–160 m$^2$.

# NEW CHARACTERISTICS | 新特色

# LUXURY, FASHION, ELEGANCE, NATURE

| HERMES Home – Townhouse of Zhixin (Middle House)

高贵 时尚 典雅 自然——《爱马士之家》置信联排别墅–中间户型

| 项目地点：中国四川省成都市 | Location: Chengdu, Sichuan, China |
| 室内设计：FRS弗瑞思空间设计有限公司 | Interior Design: Freespace design & associates Co. Ltd. |
| 设计总监/领导人：赵学强 | Principal Designer/Leader: Jack Zhao |
| 建筑面积：1 080.4 m² | Floor Area: 1,080.4 m² |
| 摄 影 师：Peter | Photographer: Peter |
| 主要材料：天然石材、乳胶漆 | Materials: natural stone, emulsion paint |

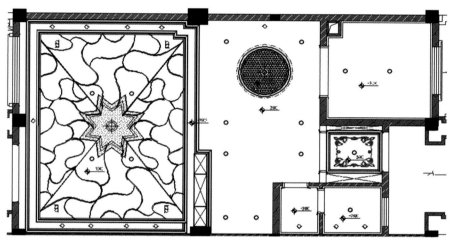

一层平面图  First Floor Plan

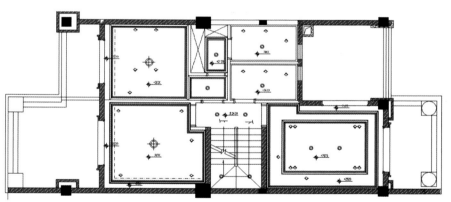

二层平面图  Second Floor Plan

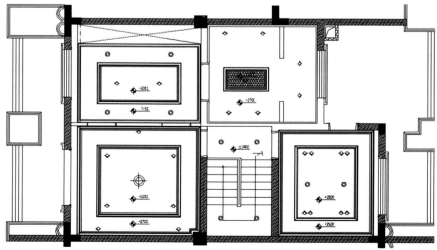

三层平面图  Third Floor Plan

这套别墅样板间作品，主要以消费群体的消费行为和心理变化为核心的设计思路和市场定位，主要针对中高端消费群体。

该中间户型，即样板间2，面积约230 m²。设计客户定位为年轻的四口之家，男主人38岁，外资银行高管，有留洋背景，深受西式文化影响，喜击剑、马术，注重生活品质，有较高的艺术修养；女主人约32岁，时尚，喜旅游、舞蹈、马术；大女儿约6岁，喜欢音乐、舞蹈；小儿子约半岁。

这套主题设计的概念来源于法国品牌爱马仕，爱马仕的产品都选用最上乘的材料，注重工艺装饰，细节精巧，以其优良的品质赢得了良好的信誉。该样板间的空间设计围绕法国中产阶级生活方式，选用壁炉、桌球台、法式贵族沙发椅等细节作为法式生活符号，并突出显示马的元素，如儿童房的木马椅、挂饰等。空间选材以天然石材和乳胶漆为主，用最少种类的材料表达完所有的空间，并且满足"时尚、环保、自然"等特点。最终营造一种故事感和情趣，以及"高贵、时尚、典雅、自然"的空间氛围。

As a showflat, the design of this house is based on the behaviors and emotions of the target buyers to meet their high-end requirements.

This 230 m² middle house is specially designed for a young family with four members. The 38-year-old host is an executive of a foreign bank. He's ever studied overseas, paying attention to the quality of life and liking fencing, riding and art. The hostess is about 32 years old, fashionable, liking travel, dancing and riding. Their daughter is about six, liking music and dancing, while their son is about half year old.

The theme of the house is inspired by the famous French brand – HERMES: all Hermes products are made of superb materials, pay attention to craft and detail, and win good reputation with high quality. In this showflat, French middle class lifestyle is shown with fireplace, pool table, luxury sofa and so on. The elements of horse are also used, such as the wooden-horse chair in Children's room and the horse-shaped hanging decorations. Natural stones and emulsion paint are used to realize a fashionable, environment-friendly and natural space. It finally creates a luxury and elegant home with story and funs.

NEW IDEA | 新创意

# WATERSIDE RESIDENCES THAT HIGHLIGHT GREEN SPACE

| Bosphorus City

强调绿色空间的滨水住宅—— 博斯普鲁斯新城

| | |
|---|---|
| 项目地点：土耳其伊斯坦布尔哈尔卡利 | Location: Halkali, Istanbul, Turkey |
| 客　　户：Sinpaş REIT | Client-Owner: Sinpaş REIT |
| 建筑设计：土耳其Mimarlar建筑师事务所 | Architectural Design: Mimarlar-Workshop Ltd |
| 建筑师：Mehpare Evrenol, Alp Evrenol, Burak Karaca, Görkem Ergazi | Architects: Mehpare Evrenol, Alp Evrenol, Burak Karaca, Görkem Ergazi |
| 占地面积：246 000 m² | Site Area: 246, 000 m² |

**项目概况与规划布局**

该项目位于哈尔卡利，它是伊斯坦布尔最新的城市改造区之一。整个地块呈东北—西南向，功能丰富的休闲区与40种不同类型的住宅楼宇呈不同方式排布其中，为伊斯坦布尔的城镇化带来了强劲动力。其中，总体规划最重要的一点是：休闲区将水与绿地完美结合，打造丰富户外休闲空间。按照地块细长的结构，水域将它一分为二，创建了一个充满活力的城市网络。绿地不是随意保留的绿色空间，而是为会议、集会、小聚等专门设计的高品质户外空间。

项目对地块东西两侧的结构密度有所加强，从而满足了滨水住宅与豪宅的功能需求。整个项目包括12层与13层不等的排房、弧形楼群、14层到20层不等的带有叶棚的塔楼。原本打算或三个或四个为一组的楼群如今被集合成一个单体，扩大了建筑物之间的距离，以加大绿地面积。

# NEW IDEA | 新创意

**Overview and Planning**

The use of waterside residences and mansions have required the structural density to increase on the west and east sides of the parcel and accordingly, 12-13 floored terrace houses, arc blocks, and towers with cascades descending from the 20th floor to the 14th can be observed. Also, triple and quadripartite blocks have been turned into a single mass to increase the distances between the buildings in order to intensify the green spaces.

This project is located in Halkali, one of the most recent urban transformation districts of Istanbul. The land plot extends in the northeast-southwest direction and provides momentum for Istanbul's urban transformation with the rich recreational areas and 40 different types of residential buildings with a wide variety of spatial arrangements. The most important element of the masterplan is the recreational areas integrating "water"

NEW IDEA | 新创意

and "green spaces" and providing enriched outdoor areas. The water, located in accordance with the long and slim structure of the land plot, divides the space into two shores and creates a dynamic urban network with inter-transits. The green spaces are not random residual spaces of green, but rather a high quality outdoor design providing possibilities for the acts of meeting / gathering / encountering. The values of Istanbul's daily urban life are protected with means of socializing in open spaces. A similar approach is also true for the residential solutions, examples of traditional civil architecture contribute a unique quality to the project in a modernized manner.

# COMME
# BUILDINGS

# RCIAL 商业地产

P126

卡伦艺术系列酒店：
"凸""凹"有致的"绿色"建筑

P132

武汉融科·天城：
融水之灵性、流畅、自然于一体的"城市客厅"

P140

温特和克希尔顿酒店：
蕴含艺术与意象之美的城市地标

COMMERCIAL BUILDINGS | 商业地产

# GREEN BUILDING WITH CANTILEVERED BOXES AND RANDOM POCKETS | The Cullen — Art Series Hotel

## "凸""凹"有致的"绿色"建筑
—— 卡伦艺术系列酒店

项目地点：澳大利亚维多利亚普拉汉商业大道
建筑设计：澳大利亚Jackson Clements Burrows建筑师事务所
项目团队：Tim Jackson, Jon Clements, Graham Burrows,
　　　　　Michael Stelluto, Rob Kennon
占地面积：1 200 m²
摄　　影：John Gollings

Location: Commercial Road, Prahran, Victoria, Australia
Architects: Jackson Clements Burrows Pty Ltd Architects
Project Team: Tim Jackson, Jon Clements, Graham Burrows,
　　　　　　　Michael Stelluto, Rob Kennon
Site Area: 1,200 m²
Photography: John Gollings

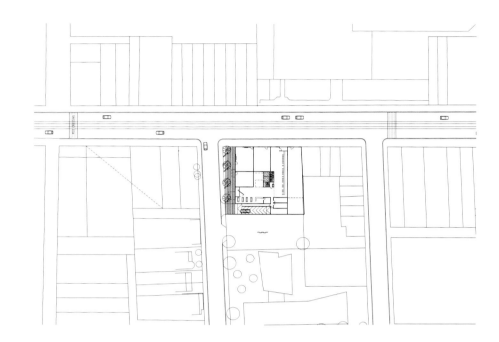

COMMERCIAL BUILDINGS | 商业地产

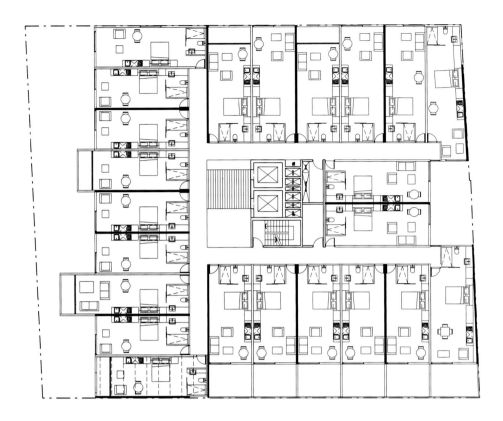

## 项目概况

这是Jackson Clements Burrows建筑师事务所受命为澳大利亚亚太集团设计的一家全新的精品酒店。整个项目包括餐厅、酒吧、咖啡厅和拥有屋顶泳池及健身馆的五层酒店。

## 建筑设计

该设计秉承的理念是要将沿Grattan花园的Greville大街与商业大道连接起来，为了实现这一想法，项目采取了两种有效的方式。首先，建筑退后Grattan大街4.5 m到6 m，留出了一个积极的公共开放空间，同时开辟了一条通往公园的视线长廊；其次，建筑通过如活雕塑一般的悬臂形式与Grattan大街相呼应，强化了这一理念。从概念上讲，这个想法是这样展开的：通过提取Grattan花园的绿意，将之点缀到建筑体量中，化成好似垂直景观的绿色悬臂盒子。面向商业大道的一面则不是伸出的悬臂盒子，而是凹进去的绿色"小口袋"。通过参考周围塔楼，设计对阳台和窗台的规模和深度都有所考究，不至于让建筑体量太过突兀，避免了违和感。

COMMERCIAL BUILDINGS | 商业地产

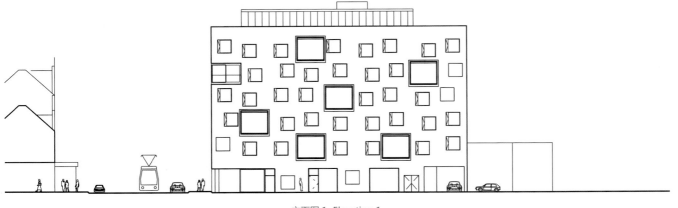
立面图 1　Elevation 1

立面图 2　Elevation 2

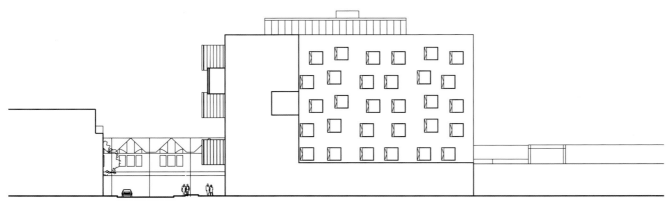
立面图 3　Elevation 3

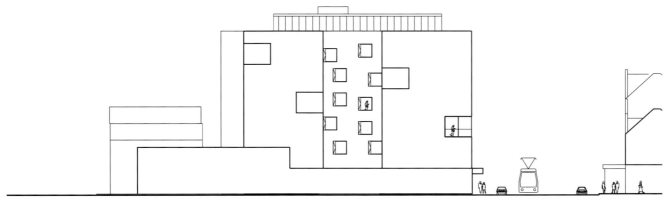
立面图 4　Elevation 4

**Overview**

JCB were commissioned to design a new boutique hotel for the Asian Pacific Group which included retail with a restaurant, bar, cafe plus five storey hotel development with setback gymnasium and pool on roof deck.

**Architectural Design**

An important idea for the architectural solution was to encourage a connection from Greville Street along the Grattan Gardens to Commercial Road. This has been done in two significant ways. Firstly, the building form is setback from the Grattan Street (4.5 – 6.0 metres) providing an active public open space while opening up a view corridor back to the park. Secondly, the built form reinforces this notion by expressing cantilevered forms as an active sculptural response to Grattan Street. Conceptually this idea was developed by abstracting the gardens beyond (Grattan Gardens) into a vertical landscape of Lichen green zinc cantilevered boxes. Internally the boxes are finished in a shade of green to reinforce the idea of a landscape that is continuous from Greville Street to the Commercial Road interface. On the Commercial road facade this idea is inverted as random 'pockets' of virtual landscape. The scale and perceived depths of the balconies and windows are manipulated, reducing the building mass and referencing the turrets and datums in the surrounding context.

# "URBAN SITTING ROOM" WITH FLEXIBILITY, SMOOTHNESS AND NATURE | Rongke Skycity, Wuhan

融水之灵性、流畅、自然于一体的"城市客厅"
—— 武汉融科·天城

项目地点：中国湖北省武汉市
委 托 方：中化方兴置业(北京)有限公司
规划/建筑设计：维思平建筑设计（WSP ARCHITECTS）
设计团队：陈凌、吴钢、隋鲁波、陈宇威、张宇、宋楠、崔广星
用地面积：155 919 m²
建筑面积：367 790 m²

Location: Wuhan, Hubei, China
Client: Sinochem Franshion Properties (Beijing) Ltd.
Planning/Architecture Design: WSP ARCHITECTS
Design Team: Chen Ling, Wu Gang, Sui Lubo, Chen Yuwei, Zhang Yu, Song Nan, Cui Guangxing
Land Area: 155, 919 m²
Floor Area: 367, 790 m²

**项目概况**

　　武汉向来以水文化著称，故该项目整体规划设计取水的流线、优美之感，充分融入现代建筑设计的语言，体现现代化、流畅、丰富的空间意境。在单体平面及商业平面多采用圆角及流线形处理，以裙房为空间造型手段，强调水平线条，犹如荷叶上扩散开来的水露，并表达"自然建筑，绿色建筑"的概念。设计中演绎发展了"客厅"作为人们购物、社交、活动之功能的场所，并延伸至城市设计的范畴，突出了该项目作为"城市之肺"的设计理念。

**规划布局与建筑设计**

　　本规划将西北侧解放南路对面的后续地块和东北侧地块统一考虑。本期地块与解放南路对面地块环形布置一栋180 m超高层酒店办公和五栋高层公寓，寓意"金，木，水，火，土"，高度分别为100 m和80 m，形成错落的天际线，超高层从地面向上扭转盘旋，犹如一条腾龙冲上云霄，结合入口广场的设计突出整体地标形象。各高层建筑基本与街道平行布置，和城市形态肌理脉络保持一致。各个高层建筑通过犹如荷叶上扩散开来的水滴形态的底层架空的两、三层商业连廊进行平面和立体的连接，形成分散但不分离的高层建筑组群，并营造较大尺度的地面和空中院落，打造区域内的"城市客厅"，作为整体项目内部一个重要节点，是内部空间的"心脏"，提供了相对开放的城市活动场所。

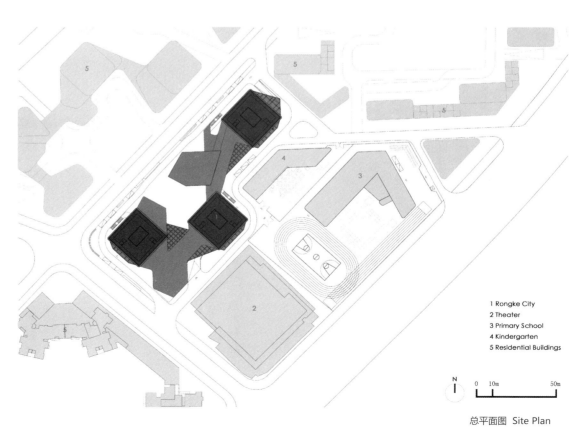

1 Rongke City
2 Theater
3 Primary School
4 Kindergarten
5 Residential Buildings

总平面图　Site Plan

COMMERCIAL BUILDINGS | 商业地产

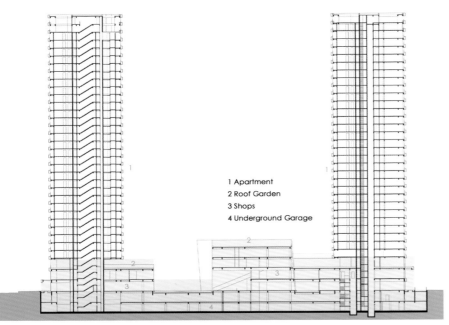

A-A 剖面图 Section A-A

1 Apartment
2 Roof Garden
3 Shops
4 Underground Garage

在中心广场布置折线性会所，犹如镶嵌水中的钻石，是对社区物业形象展示的升级。会所旁中心广场设置中心展演台、休闲座椅、茶室、凉亭等户外小品，结合绿化、水体、灯具等景观要素，形成一个尺度亲切、环境宜人的户外交流场所；一、二层商业平面好似"花瓣"点缀于流线型的商业连廊中。二层商业连廊与广场地面有三部楼梯和两部景观扶梯紧密联系。商业裙房屋顶之上是塔楼住宅的空中花园，利用群房屋顶景观架空层形成公共性交流空间。自地面而上形成多层级的景观绿化、活动空间，各功能空间相互渗透交流，作为面向城市的窗口，这将无疑会成为城市最重要和炙手可热的地方。

东北侧后续住宅地块各高层住宅亦呈环状布置，周围布置沿街商业，与西侧的中心商业广场结合形成"∞"的布局，同时将三阳路、京汉大道、解放南路、解放大道、新马路相互联系，形成汉口城市商业网络的交集。高层住宅考虑到武汉南方地区的气候特点，采用了长板式，南北朝向，确保日照通风条件的良好。

同时形成一条东西向贯穿的休闲商业街轴线。这一线性商业街将场地上的各个节点串连在一起，并将京汉大道及解放南路连接在一起。轴线的西端为后续规划的超高层建筑，形成功能、空间和视觉的通廊，同时增加了商业临街面，使商业空间更加开放。轴线东段北侧为后续住宅地块，南侧为幼儿园及学校，学校临京汉大道布置，西南侧为中南剧院，出行便利，与商业住宅互不干扰。幼儿园、学校、会所、餐饮、超市等公共设施与各建筑共同组成具有未来性的城市综合体。

## Overview

Wuhan is famous for water culture for a long time. This project is based on the stream line and beauty of water to show the modern, fluent and rich space which is integrated with modern architecture design language. Round corners and stream lines are used mostly on the monomer planes and the commercial planes. The horizontal lines are emphasized by the podiums just like the dewdrops spreading on the lotus leaf, expressing the concept of "natural architecture and green architecture". The design developed the concept of "sitting room" where shopping and social activity occur, and extended it to the urban design, highlighting the design concept of "lung of the city".

## Planning and Architectural Design

The site opposite to Jiefang South Road on the northwest and the site on the northeast are planned together. A 180-meter super-high hotel office and a 5-storey apartment are set on the site, forming the randomly stewed skyline. The super-high swirls upwards like a dragon soaring to the sky, highlighting the landmark image together with the entrance square. Most of the high buildings parallel the streets and maintain harmony with the urban texture. Every building is connected to another by the commercial corridors on the second and the third floors that are like water drops on the lotus leaf. Thus, a group of scattered but connected high buildings is formed, creating large yards both on the ground and in the air. As a significant node of the project,

# COMMERCIAL BUILDINGS  |  商业地产

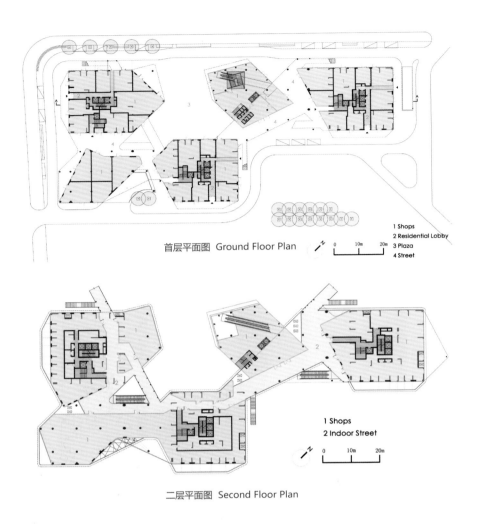

首层平面图 Ground Floor Plan
1 Shops
2 Residential Lobby
3 Plaza
4 Street

二层平面图 Second Floor Plan
1 Shops
2 Indoor Street

the "urban sitting room" is the heart of the inner space, providing open places for urban activity.

The club in the central square is like a diamond in the water, symbolizing the upgrade of the community image. Near the club, the center exhibition stage, the chairs for leisure, the teahouse, the pavilion, etc. are set together with the landscape elements such as greening, water and lamps, forming an outdoor interacting place with cordial and pleasant environment; the commercial planes on the first and the second floor are like petals embellished among the streamlined commercial corridors. The commercial corridor on the second floor and the ground of the square are connected by three stairs and two landscape escalators. Above the commercial podiums is the hanging garden where the public interaction space is formed by the landscapes. From the ground and upwards, the multiple layers of landscape and activity spaces interknit each other. As the window of toward the city, this would be he most important and popular place of the city.

The high rise residences on the northeast side are set with circle shape, surrounded by the shops and stores along the street. They form a layout of "∞" with the central commercial square on the west side and connect with Sanyang Road, Jinhan Avenue, Jiefang South Road, Jiefang Avenue and Xinma Road, forming the intersection of the commercial

network in the city. Considering the climate feature of southern part of Wuhan, the high rise residences are designed with long-plate type in north-south direction to make sure good ventilation condition.

A leisure shopping street of east-west direction is also formed, linking every nodes as well as Jinghan Avenue and Jiefang South Road. At the west end of the street, there is a super-high-rise construction with functional, spatial and visual corridor. The shops and stores along the street make the space broader. On the north side of the east part, there is the residential site. The kindergarten and the school are on the south side. The school is by the Jinhan Avenue. The theater is on the southwest side with convenient transportation. All these public facilities compose the urban complex with future sense.

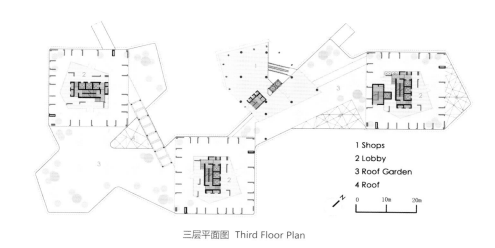

1 Shops
2 Lobby
3 Roof Garden
4 Roof

三层平面图 Third Floor Plan

COMMERCIAL BUILDINGS | 商业地产

# A LANDMARK WITH ART AND IMAGERY | Eliakim Namundjebo Plaza Hotel, Windhoek

## 蕴含艺术与意象之美的城市地标
—— 温特和克希尔顿酒店

项目地点：纳米比亚温特和克
客　　户：United Africa Group
建筑设计：纳米比亚Wasserfall Munting建筑师事务所
建筑师：Jaco Wasserfall
项目经理：Willy Klein

Location: Windhoek, Namibia
Client: United Africa Group
Architects: Wasserfall Munting Architects
Project Architect: Jaco Wasserfall
Projects Manager: Willy Klein

**项目概况**

　　该希尔顿酒店位于纳米比亚的首都温特和克。为改造城市面貌和激发中央商务区的活力，温特和克决定开发一个80 000 m²规模的综合自由广场，该酒店是整个项目的第一期。

　　出霍齐亚库塔科国际机场，沿萨姆努乔马大道至市中心，即可见酒店身影。它位于地块最南端，三面临街，周围有最高法院、纳米比亚银行和市议会，坐享城市全景和周围山景。同时，酒店可享整个发展区全貌，尤其是那条即将布置零售商店、办公和其他功能的林荫步行道。

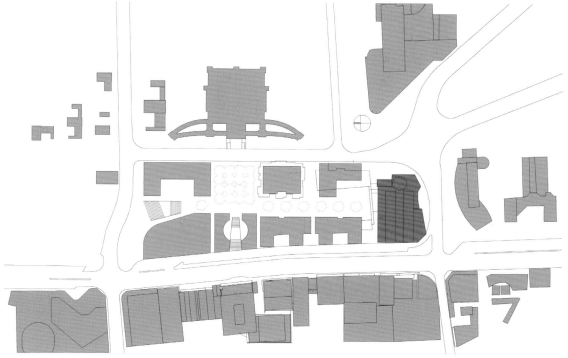

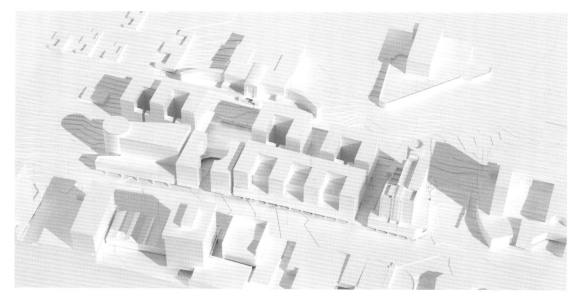

COMMERCIAL BUILDINGS | 商业地产

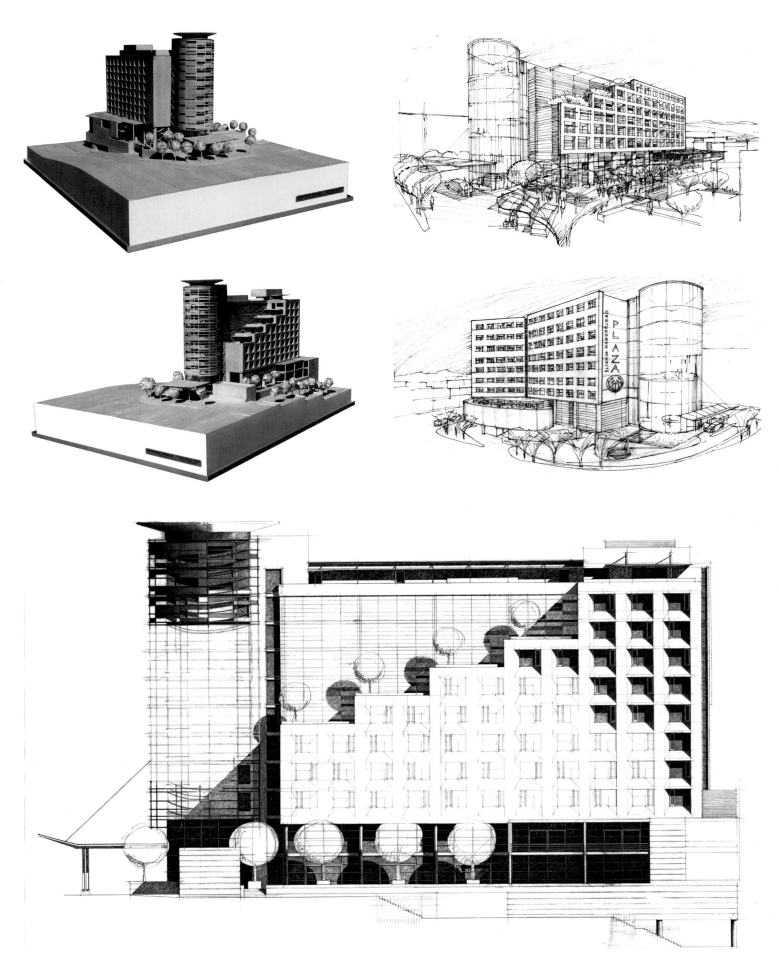

## 建筑设计

酒店正门处有一座高40 m的玻璃钢塔,迎接来客。穿过正门,接着是一个全高的中庭,从这里可以直达主楼所有的公共功能区:休息室、咖啡厅、宴会厅、会议室、全天候餐厅、酒吧和夹层露台。镶有玻璃的部分天庭引进自然光,连通了不同楼层间的视线。圆柱塔楼的最上面两层有个贵宾室,连接屋顶游泳池、池畔小屋、健身房和露台酒吧。

可识别性和透明度是设计考虑的两个重要因素。一对升降机井竖向贯通整座大楼,灯光四溢的全景电梯给它们带来勃勃生气。

希尔顿酒店被看做是数字时代城市夜生活的一盏明灯,以艺术与意象之美诠释当地文化,吸引游客的到来。希尔顿酒店之于纳米比亚正如皮卡迪利广场之于英国,时代广场之于美国。

COMMERCIAL BUILDINGS | 商业地产

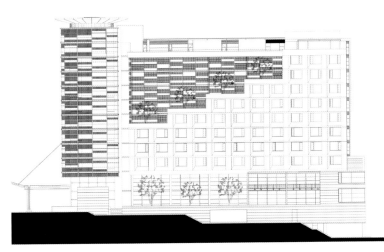

北立面图 North Elevation

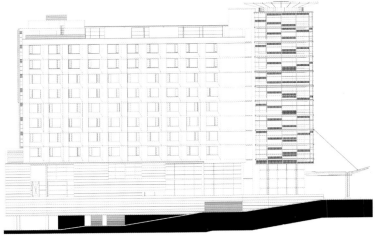

南立面图 South Elevation

## Overview

The new Hilton (Eliakim Namundjebo Plaza) Hotel in Windhoek, Namibia, forms the first phase of the proposed Freedom Square development, an 80,000 m² mixed-use precinct in the heart of the capital set to transform the urban face and dynamics of the CBD.

Prominently situated as the city centre is entered from the Hosea Kutako International Airport along Sam Nujoma Avenue, the hotel occupies the southernmost portion of the larger property and is hemmed in by streets on three sides. The elevated Supreme Court and Bank of Namibia overlook the site to the east, and the City Council Building to the south. The Hilton enjoys panoramic views of the city and surrounding mountains to the west and southwest. It will also have a commanding view of the new development

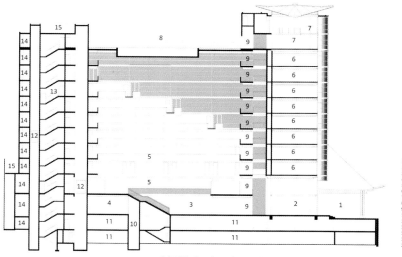

剖面图 Sectional Drawing

1. Porte Coche
2. Entrance lobby
3. Reception/lounge
4. Conference lobby
5. Atrium
6. Suite
7. Executive lounge
8. Pool deck
9. Lift lobby
10. Shuttle lift
11. Parking
12. Service lift
13. Stairwell
14. Back of House
15. Plant

# COMMERCIAL BUILDINGS | 商业地产

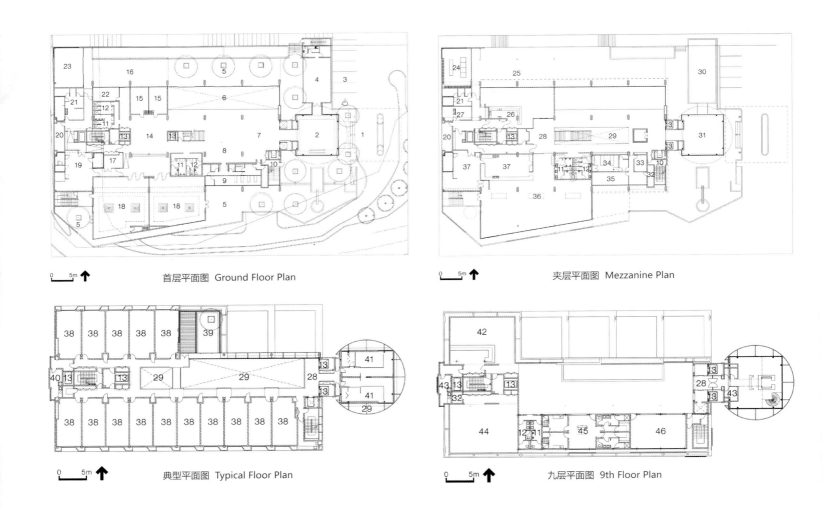

首层平面图 Ground Floor Plan

夹层平面图 Mezzanine Plan

典型平面图 Typical Floor Plan

九层平面图 9th Floor Plan

to the north and, in particular, the tree-lined pedestrian spine alongside which the retail, office, and other components will be arranged.

**Architectural Design**

Announcing the hotel to the approaching guest and marking its main entrance is a 40m high glass-and-steel tower. The prominent entrance drum links to a full-height atrium from where all public functions in the main building can be reached: the lounge and coffee shop, banqueting halls, meeting rooms and terrace on ground level, and an all-day dining restaurant, cocktail bar and terrace on mezzanine. Natural light penetrates the building interior through the partially glazed atrium that, along with other double volume spaces, creates visual links between the various levels of the building. An executive lounge is located on the upper two levels of the glazed drum and links to a rooftop pool, poolside cabanas, a gym and wellness facility, and a terrace bar.

Legibility and transparency were important design considerations. A strong circulation axis and visual links ensure clarity of public circulation, both horizontally and vertically. The idea of illuminated boxes within a legible building structure is applied as a constant theme for design elements such as retail shops, offices, bars, reception desks, and selected service components, with varying degrees of transparency reflecting the required levels of privacy. Animated by the illumination of the panoramic lift cars, the twinned main lift shafts are expressed as glazed vertical boxes penetrating the total height of the building.

Conceptually the Hilton was perceived as a beacon of urban nightlife in a digital age, celebrating arrival and local culture with art and imagery, thus providing Namibia's capital city with a public landmark comparable to Piccadilly Circus or Time Square.

COMMERCIAL BUILDINGS | 商业地产

## 深圳市佰邦建筑设计顾问有限公司 P.B.A architectural consultant co.,ltd

佰邦建筑设计公司（香港）为一家注册于香港、总部位于九龙湾億京大厦的建筑设计公司，拥有多支设计理念独到、设计经验丰富的境内外设计师团队。2006年于中国内地深圳设立了深圳市佰邦建筑设计顾问有限公司，2011年于福建福州设立了公司办事处。自公司成立以来，以积极的姿态活跃于内地设计领域，并不断的完善服务内容。 目前佰邦公司已形成一个能够就各类地产及政府项目前期决策、中期设计为一体的高品质地产服务平台，提供全方位的地产服务。公司主创设计师在各类型建筑上获得过多项国家及地方建筑设计奖项。同时公司与有实力的境外设计单位形成设计联合体，引入国际视野的设计理念，更好的为中国市场提供服务。

公司设计特色： 1、城市综合体策划和设计　　2、各类型酒店建筑设计　　3、商业建筑设计
　　　　　　　4、高端品质居住区规划和设计

深圳　Tel：+86（755)86229594　　　福州 Tel：+86（591）87524911　　　公司主页：www.pba-arch.com

# LUCAS 奥德景观
## DESIGN GROUP

西安阳光城-西西安小镇

## 深圳市奥德景观规划设计有限公司

 公司坐落于著名的蛇口湾畔,深圳最有影响力的创意设计基地:南海意库;
公司前身为深圳市卢卡斯景观设计有限公司,是由2003年成立于香港的卢卡斯联盟(香港)国际设计有限公司在世界设计之都:深圳设立的中国境内唯一公司。于2012年1月获得中华人民共和国国家旅游局正式认定:旅游规划设计乙级设计资质。

 居住区景观与规划设计(含旅游地产)
商业综合体景观与规划设计(含购物公园、写字楼及创意园区)
城市规划及空间设计
市民公园设计
酒店与渡假村景观规划与设计
旅游策划及规划设计

 倚重当下的中国的文化渊源结合世界潮流,尊重地域情感,在中国打造具强烈地域特征的、风格化的、国际化的,具前瞻性、可再生的的城市景观、人居环境、风情渡假区及自然保护区。

地　　址:深圳市南山区蛇口兴华路南海意库2栋410室
电　　话:0755-86270761
传　　真:0755-86270762
邮　　箱:lucasgroup_lucas@163.com
网　　址:www.lucas-designgroup.com